HOW I KILLED PLUTO
and Why It Had It Coming

HOW I KILLED
PLUTO

and Why It Had It Coming

MIKE BROWN

Spiegel & Grau Trade Paperbacks   New York   2012

2012 Spiegel & Grau Trade Paperback Edition

Published in the United States by Spiegel & Grau, an imprint of The Random House Publishing Group, a division of Random House, Inc., New York.

SPIEGEL & GRAU and Design is a registered trademark of Random House, Inc.

Originally published in hardcover and in slightly different form in the United States by Spiegel & Grau, an imprint of The Random House Publishing Group, a division of Random House, Inc., in 2010.

Library of Congress Cataloging-in-Publication Data
Brown, Mike.
How I killed Pluto and why it had it coming / Mike Brown.
p. cm.
ISBN 978-0-385-53110-8
eBook ISBN 978-0-385-53109-2
1. Pluto (Dwarf planet). 2. Planets. 3. Solar system. 4. Discoveries in science—Anecdotes. I. Title.
QB701.B77 2010
523.49'2—dc22   2010015074

Printed in the United States of America

www.spiegelandgrau.com

9 8 7

For Diane and Lilah

PLUTO DIES

As an astronomer, I have long had a professional aversion to waking up before dawn, preferring instead to see sunrise not as an early-morning treat, but as the signal that the end of a long night of work has come and it is finally time for overdue sleep. But in the predawn of August 25, 2006, I awoke early and was up sneaking out the door, trying not to wake my wife, Diane, or our one-year-old daughter, Lilah. I wasn't quite quiet enough. As I was closing the front door behind me, Diane called out, "Good luck, sweetie!"

I made the short drive downhill through the dark empty streets of Pasadena to the Caltech campus, where I found myself at 4:30 a.m., freshly showered, partially awake, and uncharacteristically nicely dressed, unlocking my office building to let in news crews that had been waiting outside. All of the local news affiliates were there, as well as representatives of most of the national networks. Outside, a Japanese-speaking crew was pointing a TV camera up at the sky, the beams of the flood lamps disappearing into space.

Today was the last day of the International Astronomical Union meeting in Prague, and the final item on the agenda at the end of two weeks' worth of discussion was a vote on what to do with Pluto. Everyone's favorite ice ball was in imminent danger of being cast out of the pantheon of planets by the vote of astronomers assembled half a world away, and whatever happened would be big news around the globe.

I like planets, but I didn't care enough about Pluto to get up at 4:30 a.m. But this Pluto vote mattered enough for me to drag myself out of bed that morning. For me that vote had nothing to do with the ninth planet; it was all about the tenth. And I cared a lot about that tenth planet, because eighteen months earlier, I had discovered it, a ball of ice and rock slightly larger than Pluto circling the sun every 580 years. I had been scanning the skies night after night looking for such a thing for most of a decade, and then, one morning, there it finally was.

At the time of the Pluto vote, my discovery was still officially called only by its license plate number of 2003 UB313, but to many it was known by the tongue-in-cheek nickname of Xena, and to even more it was known simply as the tenth planet. Or maybe, after today, not the tenth planet. Xena had precipitated the past year of intensive arguments about Pluto, but it was clear that Xena would share whatever fate was dealt to Pluto. If Pluto was to be a planet, then so too Xena. If Pluto was to be kicked out, Xena would get the same boot. It was worth waking up early to find out the answer.

The previous two weeks in Prague had been perhaps the most contentious gathering in modern astronomical history. Usually the International Astronomical Union meeting is nothing but a once-every-three-years chance for astronomers to advertise their latest discovery or newest idea while spending some time in a nice international destination, having dinners with old friends and catching up on their celestial gossip. On the final day of each meeting, in a session attended by almost no one, resolutions are passed, usually all but unanimously, on such pressing topics as the precise definition, to the millisecond, of Barycentric Dynamical Time (I have no idea what this actually even means).

This year was different. The usually placid astronomers had spent their time in Prague arguing and bickering day and night about Pluto and about planets. While several of the typically unintelligible resolutions were indeed to be voted on this last day, the final two resolutions would be all about Pluto. The usually sparsely attended final session was likely to be full of surly astronomers itching for a fight.

While the astronomers were gathering for their vote in Prague, the news crews and I were arriving in the early morning on the Caltech campus in Pasadena, California, so that we could watch the excitement via a webcast. My job was to provide commentary and analysis for the press and moral support and scientific coverage for the astronomers who were—rightly, I thought—trying to take the bold move of ridding the solar system of the baggage of planet Pluto. I found the webcast, projected it on the large screen, and we all sat back to watch.

Three mostly esoteric and tedious hours later, it was all over. On the final vote, the air was filled with yellow cards with which the astronomers in Prague were voting "no" to Pluto's planethood. There was no need to count; the vote was not even close. After hours of detailed explanation and analysis and discussion of the subtleties of all the different possible outcomes, I could finally just say: "Pluto is dead."

The cameras whirred; correspondents talked into their microphones, and on a screen on the other side of the room, I could see myself on some local television station repeating, like an echo, "Pluto is dead."

Before anyone else could ask a question, I quickly picked up the phone and called Diane, who was now at work. I had made a similar phone call eighteen months earlier, only minutes after I had discovered Xena. Back then, the moment she picked up the phone I said, "I found a planet!"

Back then, her voice had risen. "Really?"

Yeah! Really!

This time, instead, the moment she picked up the phone, I said, "Pluto is no longer a planet!"

Her voice dropped. "Really?"

Yeah! Really! I was still excited about the vote and had not quite grasped her mood.

She paused for a long time. "And Xena?" she finally asked quietly.

But Diane already knew the answer. Xena had indeed gotten the same boot as Pluto, and Diane was already mourning the little planet that we had gotten to know so well.

In the days that followed, I would hear from many people who were sad about Pluto. And I understood. Pluto was part of their mental landscape, the one they had constructed to organize their thinking about the solar system and their own place within it. Pluto seemed like the edge of existence. Ripping Pluto out of that landscape caused what felt like an inconceivably empty hole.

That first morning, Diane was having the same reaction, but for Xena instead of Pluto. For her, Xena was more than just that thing that used to be called "the tenth planet." She had listened to me enough over the previous eighteen months that she had gotten to know all about the onetime tenth planet. She knew about its tiny moon, its incredibly shiny surface, and its atmosphere frozen in a thin layer all around the globe. Diane and I had discussed the excitement of the search, what to name the tenth planet, and how many more like it might be out there. Xena had become as much a part of our own mental landscapes as Pluto might have been for anyone else. And Xena would be forever linked in our minds to our daughter, Lilah, who was only three weeks old when Xena was announced to the world. All of those memories of the first months of our Lilah's life—the lack of sleep, the dazed confusion, the questions about what life would be like after this sudden change—were tied up with all of our memories of what became tenth-planet mania—the rush to learn more, the push to discover others, the questions about what life would be like after this sudden change. And now, just a little after Lilah's first birthday, Xena was gone.

I had to tell Diane: The astronomers did the right thing.

Xena is not really gone, of course. It is now actually the largest of the dwarf planets, which it rightfully deserves to be.

Lilah will probably not learn about Xena in school, but someday, we'll tell her that when she was three weeks old the world first heard about the tenth planet, and we'll pull out our little box of Xena news clippings and talk about that year when Lilah and the tenth planet were both burning themselves into our lives as things that we could never again imagine the universe without.

CONTENTS

HOW I KILLED PLUTO
and Why It Had It Coming

Chapter One

WHAT IS A PLANET?

One December night in 1999, a friend and I were sitting on a mountaintop east of San Diego inside a thirteen-story-tall dome. Only a few lights illuminated the uncluttered floor of the cavernous interior, but above you could vaguely see the bottom half of the massive Hale Telescope at Palomar Observatory. The Hale Telescope was, for almost fifty years, the largest telescope in the world, but from where we sat, with the weak yellow incandescent lighting being swallowed in the darkness above, you would never have guessed where you were. You might have thought you were deep in the interior of a pristine Hoover Dam, with cables and wire and pipes for directing the flow of water around. You might have believed that the steel structures around you were part of the far underground support and control of a spotlessly clean century-old subway system. Only when the entire building gently rumbled and a tiny sliver of the starry sky appeared far over your head and the telescope began to move soundlessly and swiftly to point to some new distant object in the universe, only then would you be able to make out the shadowy outline of the truss all the way to the top of the dome and realize that you were but a dot at the base of a giant machine whose only purpose was to gather the light from a single spot beyond the sky and focus it to a tiny point just over your head.

Usually when I am working at the telescope I sit in the warm, well-lit control room, looking at computer screens showing instrument readouts, staring at digital pictures just pulled from the sky, and pondering meteorological readings and forecasts for southern California. Sometimes, though, I like to step out into the cold, dark dome and stand at the very base of the telescope and look up at the sky through the tiny open sliver high overhead and see—with my own eyes—exactly what the giant machine is looking at. This December night, however, as I was sitting with my friend inside the dark dome, there was no sky to see. The dome was fastened closed, and the telescope was idle because the entire mountain was covered in cold, dripping fog.

I tend to get quite glum on nights when I'm at a telescope with the dome closed and the precious night is slipping past. An astronomer gets to use one of these biggest telescopes only a handful of nights per year. If the night is cloudy or rainy or snowy, too bad. Your night on the telescope is simply lost, and you get to try again next year. It's hard not to think about lost time and lost discoveries as the second hand very slowly crawls through the night and your dome stays closed. Sabine—my friend—tried to cheer me up by asking about life and work, but it didn't help. I instead told her about how my father had died that spring, and how I felt unable to really focus on my work. She finally asked me if there was *anything* that I was excited about these days. I paused for a few minutes. I momentarily forgot about the freezing fog and the closed dome and the ticking clock. "I think there's another planet past Pluto," I told her.

Another planet? Such a suggestion would have generally been scoffed at by most astronomers in the last days of the twentieth century. While it is true that for much of the last century astronomers had diligently searched for a mythical "Planet X" beyond Pluto, by about 1990 they had more or less convinced themselves that all that searching in the past had been in vain; Planet X simply did not exist. Astronomers were certain that they had a pretty good inventory of what the solar system contained, of all of the planets and their moons, and of most of the comets and asteroids that circled the sun. There were certainly small asteroids still to be discovered, and occasionally a bright comet that had never been seen before would come screaming in from the far depths of space, but certainly nothing major was left out there to find. Serious discussions by serious astronomers of another planet beyond Pluto were as likely as serious discussions by serious geologists on the location of the lost continent of Atlantis. What kind of an astronomer would sit underneath one of the biggest telescopes in the world and declare, "I think there's another planet past Pluto"?

. . .

Almost a decade earlier, in the late summer of 1992, I was in the long middle years of my graduate studies at Berkeley (the place where I was taught that Planet X certainly did not exist and that we already knew pretty much everything we needed to know about what there was in the solar system). I didn't think much about Planet X those days. I was midway through a Ph.D. dissertation about the planet Jupiter and its volcanic moon Io. When you're midway through a Ph.D. dissertation, your mind acquires narrow blinders, so I didn't think much about *anything* other than Io and how its volcanoes spewed material into space and affected Jupiter's powerful magnetic field. I had so few thoughts to spare for most of the quotidian universe that I had fallen into a pattern of every day eating the same lunch at the same coffee shop right next to the Berkeley campus and having dinner at the same burrito stand a block away. At night I would ride my bicycle back down toward the San Francisco Bay to the marina where I lived on a tiny sailboat. The next morning I would start all over again. Less time thinking about what and where to eat and sleep meant more time thinking about Io and volcanoes and Jupiter and how they all fit together.

But, occasionally, even obsessive Ph.D. students need a break.

One afternoon, as on many times previous, after spending too much time staring at data on my computer screen and reading technical papers in dense journals and writing down thoughts and ideas in my black bound notebooks, I opened the door of my little graduate student office on the roof of the astronomy building, stepped into the enclosed rooftop courtyard, and climbed the metal stairs that went to the very top of the roof to an open balcony. As I stared at the San Francisco Bay laid out in front of me, trying to pull my head back down to the earth by watching the boats blowing across the water, Jane Luu, a friend and researcher in the astronomy department who had an office across the rooftop courtyard, clunked up the metal stairs and looked out across the water in the same direction I was staring. Softly and conspiratorially she said, "Nobody knows it yet, but we just found the Kuiper belt."

I could tell that she knew she was onto something big, could sense her excitement, and I was flattered that here she was telling me this astounding information that no one else knew.

"Wow," I said. "What's the Kuiper belt?"

It's funny today to think that I had no idea what she was talking about. Today if you sat next to me on an airplane and asked about the Kuiper belt, I might talk for hours about the region of space beyond Neptune where vast numbers of small icy objects circle the sun in cold storage and about how, occasionally, one of them comes plummeting into the inner part of the solar system to light up the skies like a comet. I might talk about the very edge of the solar system, where millions of little icy bodies never quite got gathered up into one big planet but instead stayed strewn in the disk surrounding the solar system. And I might tell you a little history, about how in the early 1990s no one had seen such a thing as this Kuiper belt, but a small group of astronomers who had predicted its existence had named the region the Kuiper belt after Dutch American astronomer Gerard Kuiper, who had speculated about its existence decades earlier. And finally, if you were still listening and the plane had not yet landed, I would tell you how this Kuiper belt was finally seen, for the first time, in the late summer of 1992, and how I first learned about it on the roof of the Berkeley astronomy building a day before it appeared on the front page of *The New York Times.*

But when Jane told me she had just found the Kuiper belt, I didn't know any of this. Jane explained. She had not found this vast collection of bodies beyond Neptune, exactly, but simply a single small icy body circling the sun well beyond the orbit of Pluto. The body was tiny—much, much smaller than Pluto—and as far as anyone knew for sure, it might have circled the sun all alone at the edge of the solar system. But still, exciting, right?

Cute, I thought. But it's just one tiny object, and it's farther away than Pluto. How could that matter?

So I nodded and listened and, like any diligent Ph.D. student midway through a dissertation, eventually walked back down the stairs, stepped into my office, and reentered the world of Jupiter and Io and volcanoes, where I actually resided.

I was wrong, of course. Even though the object discovered was only a lonely, relatively tiny ball of ice orbiting beyond Pluto, it showed that astronomers had been wrong: They didn't actually know everything; there were things still to be found at the edge of our own solar system. Some astronomers were reluctant to consider this new possibility seriously, and they dismissed the discovery as nothing more than a fluke that presaged absolutely nothing. But soon, as more and more astronomers became excited about the possibility of discovery and started searching the regions beyond Pluto, more and more of these small bodies began to be found.

By the end of 1999, on the foggy December night when Sabine and I were sitting underneath the Hale Telescope at Palomar Observatory and I was proclaiming that I thought there were new planets to be found, astronomers around the world had already discovered almost five hundred of these bodies in a vast disk beyond the orbit of Neptune in what looked very much indeed like the Kuiper belt. From being something that most astronomers had perhaps heard of once or twice, the Kuiper belt had become the hottest new field of study within the solar system.

Of the five hundred bodies that were then known in the Kuiper belt in 1999, most were relatively small, maybe a few hundred miles across, but a few moderately large objects had also been found. The largest known at the time was somewhere around a third the size of Pluto. A third the size of Pluto! Pluto had always enjoyed a somewhat mythical status as a lonely oddball at the edge of the solar system, but it turned out that it had more company than astronomers had originally thought.

Over the years since I had dismissed the entire Kuiper belt as not quite interesting enough to pull my mind away from Jupiter, I *had* actually been thinking a bit about Pluto and about those five hundred small icy bodies recently discovered in the distant solar system. By now it seemed to me inevitable that, whether anyone realized it or not, astronomers were on an unstoppable march that would eventually lead to a tenth planet. It seemed to me obvious that it was there, slowly circling the sun, just waiting for the moment when someone somewhere pointed a telescope at the right spot, noticed something that hadn't been there earlier, and suddenly announced to an unsuspecting world that our solar system had more than nine planets.

Sitting beneath the massive Hale Telescope that foggy night, ever the scientist, Sabine asked, "What evidence do you have?"

I told her about all of the recent astronomical discoveries. But when pressed for evidence, I had to admit: I had none whatsoever. I had a hunch. Officially, scientists don't work on hunches. We work on hypotheses and observations and plenty of evidence. Hunches don't get you research funding, tenure at your university, or access to the world's largest telescopes. But a hunch was all I had. No one had systematically looked across the sky for a new planet since the 1930s, when Pluto itself was found, and even though astronomers knew of almost five hundred bodies in the Kuiper belt, the searches had been, of necessity, piecemeal, and no one had yet mounted a careful search like the one that had uncovered Pluto. Now, seventy years after the discovery of Pluto, telescopes were bigger and better, computers made searches vastly more powerful, and astronomers simply knew more about what they were looking for. How could it be that if someone went and looked again for a new planet they wouldn't find something that had been just beyond the reach of the telescopes in the 1930s? There *had* to be a tenth planet. The possibility that Pluto was a unique planetary oddball out at the edge of the solar system seemed absurd to me.

"I don't have any evidence," I told Sabine. "I don't have any proof. I don't have anything other than this deep feeling that another planet past Pluto makes sense. And I'm willing to bet that there's one there."

Scientists don't bet much. We are supposed to deal in quantifiable levels of certainty and in statements that can be backed up with experiments and observations. Bets are simple assertions that you think you are right and that you believe what you are saying enough to risk something valuable if you're wrong. There is nothing scientific about a bet at all; in fact, it is almost the opposite of science. In earlier years many scientists would have bet the farm against the big bang, evolution, and quantum mechanics, and the farm would be gone.

But still, there's something appealing about betting. I had no solid evidence to go on, but bits and pieces of different facts and discoveries had, somehow, shaken together in my mind to form a hunch. Though I couldn't prove it to a scientist, I was all but certain that I was correct. I couldn't prove it, but I could definitely bet on it.

Sabine took the wager. The bet was that someone would find a new planet by December 31, 2004. The winner of the bet would receive five bottles of champagne, to be drunk in celebration of new planetary frontiers or in mourning for the sad limitations of our solar system.

We sat for a few minutes staring up at the telescope, thinking about planets.

"We've got one problem. We'll never know if someone wins the bet," I said.

"What?" she asked. "How could we not know whether or not someone finds a planet? Surely the entire world will hear about it. It'll be pretty obvious."

"Well, okay," I said, "then I have one question for you: What is a planet?"

I needed to know the answer, because I wanted to find one myself.

. . .

Like most everyone else, I've known what a planet is since I was four or five years old, which for me would have been about 1970. I knew the moon even earlier. I grew up in Huntsville, Alabama, a thoroughly dedicated rocket town. The father of everyone I knew—mine included—was some sort of engineer working to build the Apollo rockets to send men to the moon. For a while as a child, I thought that when you grew up you became a rocket engineer if you were a boy and you married a rocket engineer if you were a girl; few other options in the world appeared to exist. When Neil Armstrong stepped on the moon, I was pretty sure that that was exactly what I was going to be doing eventually, too. I drew picture books of rockets blasting off, of command capsules in orbit about the moon, of lunar modules landing next to giant lunar craters, and of parachutes deployed in the moments before splashdown.

By second grade I had learned enough about the moon to know that those giant craters I had been drawing earlier had been formed by meteors slamming into the moon's surface. I figured out that if I went to the backyard and soaked the deep red dirt with a hose, I could throw rocks from above and make the mud look just like the craters on the moon. I could even throw the rocks sideways into the mud and make oblong craters like some I had seen in lunar pictures.

Though the moon was my favorite, I learned about planets, too. But planets were a little more abstract than the moon, since you couldn't see them and no one had stepped on them or taken pictures from the surface. Still, by first grade I had a poster on my bedroom wall that showed the solar system with an artist's conception of each of the planets. Though I didn't realize it at the time, spacecraft had already visited Mars and Venus and Mercury, so some of the pictures were quite detailed. (I didn't know about these spacecraft at the time because Huntsville was totally dedicated to the Apollo rocket program and the moon, as far as I could tell. The robotic exploration of the other planets was being run out of a small town I'd never heard of called Pasadena, on the other side of the continent.) On my poster, Mercury looked much like the moon, battered by meteors. Venus appeared only as a swirl of clouds. Mars had giant volcanoes and deep canyons. In the outer part of the solar system, things on my poster got fuzzier, since truly no one had ever seen them except through powerful telescopes; but Jupiter had its clouds and great red spot, Saturn had its rings, and Uranus and Neptune had their retinue of moons. Pluto, however, was the most exciting of them all, because it was so different from all of the other planets.

Even as a first grader I could see that Pluto didn't travel in perfect circles around the sun the way the other planets did. I could see on the poster that it came close enough to the sun to momentarily pass inside the orbit of Neptune, but the poster showed only this inner bit of Pluto's orbit. The outer parts of the orbit were so far away that Pluto would have to travel off my poster, onto my wall, out my window, and midway across the front yard toward the street before it turned around and came back in toward the sun. Even stranger, Pluto didn't orbit the sun in the same nice flat disk that all of the other planets did: It was tilted away from the others by almost twenty degrees. On the poster, all of the other planets were represented by paintings of a global view of the surface seen from high above, but Pluto—only special Pluto—had a painting of what the planet would look like if you were standing on the surface looking back at the tiny dim Sun. The surface of the planet was covered in icy spires. These days I realize that the artists would have had no idea what Pluto looked like and probably felt the need to make the surface look like *something* interesting, but as a first grader I was thoroughly convinced that Pluto was covered in icy spires and that they would shatter at the slightest touch by a future Neil Armstrong. Clearly Pluto was different and mysterious and potentially very fragile. It would take another thirty-five years for me to learn just how fragile it really was.

In third grade we finally learned about planets in school. Most people I know memorized their order by learning some variant of the mnemonic "My very excellent mother just served us nine pizzas" for "Mercury Venus Earth Mars Jupiter Saturn Uranus Neptune Pluto," but for some reason, in my school we learned one that I have never heard since: "Martha visits every Monday and just stays until noon. Period." The "and" appears between Mars and Jupiter, just where the asteroids are, though I always suspected that that was just dumb luck. The "period" at the end, though, seemed fishy even in third grade. It didn't seem so much as to make Pluto special, as the other odd characteristics did, as much as to make Pluto seem an afterthought or a late addition or just perhaps an undesirable misfit.

Oddly, though, for a kid interested in planets, I had never been very interested in the actual night sky. Sure, I could name some of the more obvious constellations and sights—the Big Dipper, Orion, the North Star. I could point out the Milky Way galaxy, which was actually visible in the dark skies above Alabama, and I could even convince the other kids that it really was the Milky Way they were seeing and not just clouds in the sky as they always seemed to think. Once, I even saw a real comet through binoculars when my father dragged me out of bed one cold winter night in 1973 and drove us to the top of a dark mountain to see what was supposed to be the spectacular Comet Kohoutek but instead looked to me like a shaky little smudge and please could I go back to sleep now? But I was never one of those kids who built his own telescope by grinding mirrors from blanks or who memorized the locations of each of the nebulae hidden among the constellations or, even, who could tell you that the bright light above the just-set sun was, in fact, not an airplane but the planet Venus. I could passionately describe the rings of Saturn, the number of moons of Jupiter, the rocky plains of Mars, and, of course, the icy spires on Pluto, but the fact that these distant worlds were up in the sky above me was never really part of how I thought about them, much like when I think of Antarctica now I think of pictures and descriptions and maps, but I never really think about the fact that if I jumped in a boat, turned south, and started sailing, I would actually end up there.

I did get a telescope for Christmas when I was in the third grade—the seemingly perfect gift for a kid like me—but I could never make it work. My brother was capable of constructing elaborate LEGO structures for any purpose and could make balsa wood airplanes that looked sleek and flew straight and were painted beautifully. I was lucky if my LEGO constructions stayed together and were made of more or less the same colors. My attempts at balsa airplanes usually ended in my deciding that, really, I had meant to make that model of an airplane *wreck,* and yes, it would be fun to burn the whole thing now. Trying to make the telescope work went little better. I needed to carefully align mirrors and keep the tripod steady and adjust eyepieces, and it never worked. I think I found a single star once— though in retrospect, knowing now what a star *should* have looked like in such a small telescope, it is entirely possible that I only looked at an out-of-focus streetlight with a shaky telescope.

One night in the late fall when I was fifteen years old, I was awake late enough to find myself looking up at Orion—the one truly familiar part of my winter night sky—and I noticed that something didn't look right. Orion is full of bright stars that make very clear patterns even for the casual sky glancer: three stars for the belt, a dagger beneath, and a quartet of bright stars outlining the rest of the body. They are among the brightest stars in their region of the sky and nearly impossible not to recognize. And yet somehow, overhead, a little to the left, there was a pair of stars every bit as bright as those of Orion that I didn't recall ever having seen before. I was not a photographic-memory-star-pattern-recognizing kid and just assumed I had somehow overlooked them, much the way I would also overlook my allegedly lost shoes even when they were right in the middle of the floor in my room. As the months went on, however, the two stars did something extraordinary. They moved! You would have never noticed it in a single night or even in a single week. But over the months, they very slowly crawled closer together. As the winter wore on and moved into spring, the two then moved apart and then around each other in an elaborate dance high overhead, while the remainder of the stars remained fixed in their constellations. I found myself eager to go check on the stars night after night. In the winter, I would have to stay up late before they rose in the sky, but as spring came, the dancing stars were directly overhead as soon as the sun went down.

I didn't ask or talk to anyone about the moving stars; I just silently kept track. At some point that spring, though, I came across a single-paragraph article in the newspaper describing the once-every-twenty-years close conjunction of the two largest planets, Jupiter and Saturn, which looked like two bright stars wandering near Orion. They were planets! Today, I am surprised that I could possibly have been as shocked as I was. How could

I not have known? What did I think those moving stars were? How at fifteen could I have seen something unknown in the sky and not immediately needed to know what it was?

I guess no one had ever mentioned to me that you could actually see planets in the sky overhead. As soon as I realized that my two moving stars were Jupiter and Saturn, however, it became clear: Planets were not just an artist's conception on my poster, nor even just images sent from distant spacecraft, but they were bright points of light that moved among the stars. Imagine how you might feel if you had been looking at pictures of the Grand Canyon all of your life and passionately studying the layered geology and tracing Powell's trip down the canyon on the first raft expedition on a topo map, and then, suddenly, while out on what was supposed to be an ordinary afternoon stroll, you turned a corner and came unexpectedly to the canyon rim and almost fell in. At that point, how could you not want to explore every corner, every tributary, and learn everything that you could possibly learn about this wonder in your own backyard?

I have been hooked on the real planets in the sky ever since. I've kept track of Jupiter and Saturn in their travels through the stars season after season. Each year they move a little farther east in the sky as they orbit around the sun. Saturn is so far away and moves so slowly that it takes a full thirty years to complete one orbit. Today, almost thirty years after I first noticed Saturn above, it has finally almost completed one of its transits all the way around the sky—one full Saturn year—and when I look outside at night I see that it is almost back in the same place where I first saw it when I was a teen wondering what those bright stars were that danced. With luck, I'll get to watch Saturn trek all the way around the sky and end up in this spot once more in my lifetime, but probably not twice.

Jupiter, closer to the sun, is comparatively fast; it takes only twelve years to go completely around the sky. When it gets to where it started, though, Saturn has moved on. It takes another eight years—twenty years in total—for Jupiter to finally catch up to Saturn once again so they come close together in a conjunction just like the one I noticed when I was fifteen. I've often wondered about the timing of this conjunction. If I had been born a few years earlier, I would have looked up at age fifteen, but Jupiter would not yet have caught up to Saturn's position in the sky. I would have noticed only one bright planet moving a little below Orion instead of two. Would I have noticed their dance? Would I have become the person I am today, someone whose first instinct when walking outside at night is to always look up, check the stars, look for planets, locate the moon? It's impossible to know, but it's always hard not to feel that in some ways, for me at least, perhaps the early astrologers were right: Perhaps my fate actually *was* determined by the positions of the planets at the moment of my birth.

Whether or not the planets controlled my fate, one thing was clear: I knew what a planet was. As a child, I knew planets from my poster on the wall. As a teen, I knew them from watching them move across the sky. And later I knew them from years of writing a Ph.D. dissertation. Nobody was going to be able to change my mind about what a planet was. Right? So then, as my friend Sabine and I were sitting inside the Hale Telescope dome at Palomar Observatory on a cloudy, drizzly night finalizing our bet about whether or not someone would find a new planet, why was it that astronomers around the world suddenly could no longer agree on a definition of the word *planet*? How could it be that even I was unsure about what would and would not count?

Chapter Two

A MILLENNIUM
OF PLANETS

The end of the twentieth century was not actually the first time that the word *planet* had become confusing. The word has existed for thousands of years, and its meaning has been continually updated to reflect our continually shifting view of the cosmos. Over the millennia there have been a few major events leading to dramatic changes.

The original ancient Greek meaning of the word *planet* was simply "wanderer," or something that moved in the sky. When, as a teenager, I first noticed Jupiter and Saturn dancing among the stars, I was seeing the sky as it had been seen for millennia and noticing that there were things that were special, things that stuck out, things that moved in a different way. As the sky slowly revolves throughout the year, the stars stay in fixed patterns while the wanderers move separately and conspicuously through the constellations of the zodiac. The ancient Greeks and Romans knew seven wanderers in the sky: the five visible planets— Mercury, Venus, Mars, Jupiter, and Saturn, which are all easy to see if you know where and when to look—plus the moon and the sun, which both also move through the sky and were also considered planets in good standing.

In a pre-electric-light, pre-urban world, people must have been much more intimately connected with the sky and the planets. Mercury and Venus, which are close to the sun and thus only show up low in the early-evening or early-morning sky, are these days frequently mistaken for airplanes; even I sometimes mistake them. But before we became used to the idea of artificial lights in the sky, the recurring appearance of the evening or morning star would have been an obvious and spectacular event that would have been hard to miss. Mars, distinctly red in the sky, even to the naked eye, always stands out. It is no wonder that some of the earliest recorded scientific records of any sort are of the positions of the planets. Everyone would have known what a planet was back then. Planets mattered. And it is no wonder that all of our basic units of time are based on the sky: A year traced the time it took for the sun to go all the way around the sky to reappear at the same location again, while a month ("moon"-th) is about the amount of time it takes for the moon to circle the earth. The seven days of the week are even named after the seven original planets. Sunday, Mo[o]nday, and Satur[n]day are the most obvious, while Tuesday through Friday are more than a bit obscure. Tiw was an ancient Germanic god of war, as Mars was to the Romans, so Tuesday is actually Mars's day. Wednesday is Woden's day. Woden was the carrier of the dead—a Germanic grim reaper—fulfilling one of Mercury's less well known jobs. Thor was the Norse god of thunder, like Jupiter, and Friday is the day of Venus in the guise of the Norse Frigga, the goddess of married love.

Though planets were so deeply embedded into many aspects of everyday life, there is no recording of the public reaction to the first and most significant shock to the word *planet*. In the six-teenth century the idea began to spread that the sun, rather than the earth, was at the center of the universe and that the earth and the planets revolved around it. Suddenly, the wander-ers were in disarray. Instead of the sun and the moon and the other planets revolving around the earth, five of them (*the plan-ets*) went around one of them (*the sun*), while the seventh (*the moon*) went around the earth. The earth, like five of the wander-ers, also went around the sun. Copernicus wrote down what is perhaps the most startling proposition of all time: "The motions which seem to us proper to the Sun do not arise from it, but from the Earth and our orb, with which we revolve around the sun like any other planet." We revolve around the sun like any other planet! The sun doesn't move; the earth does. The earth under our feet is like any other planet in the sky. The earth *is* a planet. What seems so obvious and ingrained in us today must have been profoundly disorienting. I've tried to put myself in the frame of mind of the time and tried to understand how shock-ing it would have been, but I've never come close. It is as hard for me to image an Earth-centered universe as it would have been for them to imagine anything else. Everybody thought they knew what a planet was, and suddenly, one appeared beneath their feet.

And what of the moon? At least Earth was special in that, of all the planets, it alone had another body going around it. But when Galileo first pointed his crude telescope at the sky in 1609, he discovered that Jupiter, too, had objects going around it (now called the Galilean satellites). Any reasonable pair of binoculars will show you the same thing. Find Jupiter, point the binoculars up (lean against a wall to steady your shaky hands), and you'll see the disk of Jupiter and maybe even some bands of wispy clouds on the disk. Perhaps you'll also see four tiny white dots strung in a line all to one side of Jupiter. The next night look again, and one of the tiny dots might be missing—hidden behind Jupiter—and one might have moved to the other side. The next night they will change again. The little moons are wandering around the wanderer. One of them even has volcanoes. I could tell you a lot about those volcanoes.

Even with Galileo's primitive telescope, he could tell that there were stars in the sky that were too faint for the eye to see. Did he or anyone else think about whether or not there were planets in the sky that were too faint for the eye to see? No one appears to have written about the possibility. Perhaps no one even thought about it. Though the planets had been rearranged and now were secondary to the sun, and the earth had been demoted from the center of the universe to the same status as the other planets, perhaps the possibility of additional planets circling the sun so faintly that we wouldn't know about them was simply beyond comprehension. Why would such invisible things have been put there in the first place?

It took almost two more centuries to stumble upon the answer. In 1781 the British astronomer William Herschel was charting faint stars that could be seen only through his new advanced telescope. He came to one star that looked bigger than the stars around it, which was strange, since all of the stars look simply like points of light and none appears bigger than another. When he looked again the next night, the unusual star had moved. He had found a new wanderer. But since it couldn't be a planet (obviously, since all of the planets were known, right?), what was it? Herschel assumed it was a comet near the earth. Within only a few months, however, he realized that the new object was in a circular orbit well beyond Saturn, where nothing else had ever been seen before. It was no comet, it was a planet. Herschel measured the size of the greenish disk he had found and realized that this new body had to be big—not quite as big as Jupiter or Saturn, but much bigger than any of the other planets in the solar system. The word *planet* quite naturally expanded to include this new body distantly circling the sun: the seventh planet had been found. Jupiter, the largest planet, was named after the king of the gods. Saturn, originally the most distant known planet, was named after the father of Jupiter. The new wanderer, even more distant than Saturn and unrecognized throughout history until the moment Herschel distinguished it from the stars around it, was—eventually, after sixty years of debate—named Uranus, for the most ancient of all the gods. The element uranium, discovered only seven years later, was named in honor of the new planet.

Everybody had known there were only six planets until the moment the seventh was found, but once the prejudice against the idea of new planets was overcome, the idea that there could be other unseen planets was infectious, and as the techniques to build telescopes became more and more available, people began to systematically search the skies for new wanderers. Success came more quickly than expected. On the first day of 1801, Italian astronomer Giuseppe Piazzi—who, like Herschel, had been busy studying stars, not wanderers—discovered the new planet Ceres, the eighth planet, orbiting between Mars and Jupiter.

The eighth planet? Ceres? Most people today have never heard of "planet Ceres," but there was little question at the time that Ceres was indeed a planet. Within a few years it could be found in all astronomy textbooks, alongside Uranus and the others. In keeping with tradition, the element cerium, discovered two years later, was named for the new planet. Most people today have never heard of the element cerium, either, but it is used in the walls of most self-cleaning ovens.

Planet Ceres's problems began just a year after its discovery, when German astronomer Heinrich Olbers, investigating the new planet with his telescope, accidentally stumbled upon yet another unknown object wandering through the sky: the ninth planet, Pallas! Again, there was little question that Pallas was the ninth planet, and the element palladium was named for it in 1803.

Ceres and Pallas, though considered full-fledged planets, had a few puzzling properties. While all of the other planets were well spaced in their orbits around the sun, Ceres and Pallas were, in the cosmic scheme of things, right on top of each other between Mars and Jupiter. They were different from the other planets in other ways, too. The recently discovered Uranus was too faint to be seen without a telescope simply because it was so far past Saturn. With the aid of a telescope, though, the green outline of the disk of Uranus was apparent. But Ceres and Pallas were closer to us than Jupiter, closer than Saturn. They could not be seen without the aid of a telescope not because they were far away, but simply because they were so small compared to all of the other planets. They were so small, in fact, that even with the best telescopes of the day they just looked like little points of light. Herschel, the discoverer of Uranus—wanting to preserve the uniqueness of his own discovery, I suspect—coined the term *asteroid* ("aster" is Greek for "star," as in *astronomy,* while "oid" means "like") to describe these new objects. To Herschel, Ceres and Pallas weren't like real planets with their visible disks; they appeared "starlike" instead.

Astronomers quickly found two more planets in this same region between Mars and Jupiter—the tenth planet, Juno, in 1804 and the eleventh planet, Vesta, in 1807—and then, for almost forty years, nothing new came along. This was too many new planets for some people, chemists in particular. There are no elements named after Juno or Vesta. But still, forty years was long enough for the eleven-planet solar system to firmly emplace itself into the teachings of the day. In a secondary school text from 1837, the chapter between "The fourth planet, Mars" and "The ninth planet, Jupiter" is simply called "The fifth, sixth, seventh, and eighth planets." The schoolkids who had learned about the eleven planets were probably unhappy with what was about to come.

I have never seen these fifth, sixth, seventh, or eighth planets, even though they are just as easy to see in my binoculars as the satellites of Jupiter, and I look at the satellites of Jupiter in my binoculars all the time. In fact, I love a solar system tour with good binoculars. The rings of Saturn pop out, as does the redness of Mars, and sometimes even the little crescent-moon-shaped sliver of Venus that proved to Galileo that Venus orbits the sun. I can explore the craters and mountains and shadows on the moon for hours. I've carefully tracked down the position of Uranus and stared at it several nights in a row just to experience how Herschel might have felt about his discovery. But I've never even thought to look for any of these objects that were the most exciting astronomical discoveries of the early nineteenth century.

The reason I've never looked for these four individuals, I think, is that just as the four new small planets were becoming accepted as part of our understanding of the universe, a deluge of new objects started to be discovered. By 1851, fifteen more of the new asteroid planets were found, as well as one more large planet—Neptune. Neptune was even deemed large and important enough to name an element, neptunium, in its honor, but almost no one can recall the names of the other fifteen. It was a confusing time. What counted? What didn't? On the wall in my office at Caltech I have a collection of maps of the solar system dating from about 1850 to 1900. Each map labels the solar system differently. A page from a sky atlas drawn in 1857 clearly shows Ceres, Pallas, Juno, and Vesta as "small planets," while dozens of other asteroids are generally shown in the "zone of asteroids" between Mars and Jupiter. A German map from a year earlier lists all of the known *Asteroiden* by date of discovery, with no references to their being planets at all. Even by 1896, the solar system map from the Rand McNally Atlas explicitly states that the solar system contains only the sun, planets, and comets—asteroids are not mentioned at all—and that planets are either primary (what we would call planets today) or secondary (what we would call moons). In the margins of my Rand McNally map are drawings of how big the sun would look from the planets. At the top of the margin, the sun, seen from Mercury, is huge. At the bottom, the view from Neptune shows a tiny, distant disk. In the middle are the views from Ceres, Pallas, Juno, and Vesta, still tenuously holding on to their claims to be planets. The sun looks exactly the same from each of these four since they are all the same distance away.

By the turn of the century, though, somehow all of the confusion about what was and wasn't a planet had settled. I cannot find anything written or drawn in this period that doesn't separate the asteroids from the planets. What was their offense that they were cast down from the pantheon? In the end, their major sin seems to have been that there were too many of them all in the same place. The big planets go around the sun in orbits far from one another with no overlap, but the hundreds of known asteroids had crossing and overlapping orbits and were all one big jumble. How many is too many? When there were only four and the solar system appeared stable at eleven planets—which it did for forty years—no one (except the chemists, who couldn't discover elements fast enough) seems to have complained. But the prospect of a never-ending parade of smaller and smaller planets all in essentially the same orbit around the sun was too much. There was no official vote or pronouncement, but by the early 1900s it became conventionally agreed that the solar system had only eight planets. Planet Ceres, which had held on for a century, along with all of its smaller neighbors, was demoted, with no outcry from the citizens of planet Earth.

By recognizing that Ceres and the swarm of other new bodies were fundamentally different from planets and should be classified differently, astronomers had—perhaps inadvertently, but certainly profoundly—changed the scientific meaning of the word *planet*. The word no longer simply meant anything that moved around the sun and wandered around the sky. Asteroids wandered, but they wandered in a swarm; they were the schools of minnows swimming among the pod of whales. Planets were the whales of the solar system.

As a kid I knew asteroids, too. On my poster on the wall they looked like tiny pebbles strewn in a vast band between Mars and Jupiter. They were the things—the meteors—that sometimes hit the moon and made those giant craters. I had seen shooting stars, which I knew were tiny fragments of these asteroids burning up in the earth's atmosphere. Maybe I didn't know their individual names or anything specific about them, and perhaps as individuals they were indistinguishable. But from what I knew by the time of my 1970s childhood, the difference between a planet and an asteroid was as obvious as the difference between a boulder and a handful of sand.

After the uncertainty and confusion about planets had been settled for a few decades and textbooks were clear that there were eight and only eight planets, the ninth was finally discovered. Clyde Tombaugh found Pluto by taking repeated pictures of the sky and comparing them to see if anything had changed. On February 18, 1930, he found a faint object that moved from one night to the next: a new wanderer! Unlike the myriad asteroids (hundreds were known by then), Pluto was not between Mars and Jupiter; it was well beyond Neptune, where a real ninth planet should be. Still, it was a bit odd. It was found to go around the sun in an elongated, rather than circular, orbit, and while all of the planets orbit the sun in a flat disk, Pluto was found to be tilted by almost twenty degrees away from the rest. Pluto also looked different. It was so small that you couldn't tell it was a planet at all. In fact, it appeared starlike. Some astronomers didn't want to call Pluto a planet. Shouldn't it just be called an asteroid instead? By then, though, the word *asteroid* had lost its literal meaning of "starlike" and instead referred specifically to that belt of objects between Mars and Jupiter. Should it be called a comet? Comets can have elongated and tilted orbits like Pluto's, but none had ever been seen so far away, and the word *comet* (from *coma,* Latin for "hair") specifically refers to the fuzzy appearance of comets in the sky. Pluto was not fuzzy; it looked like a star, albeit one that moved. Though it looked and behaved like no other planet known, there was no other way to classify it, so it became accepted as the ninth planet, had the element plutonium named for it, and remained unchallenged for almost seventy years as the tiny lonely oddball at the edge of the solar system, the planet with the ice spires, the planet with the orbit so extreme that it couldn't even fit on my poster on the wall, the incongruous period at the end of the solar system.

What I didn't immediately grasp when Jane Luu joined me on the roof overlooking the San Francisco Bay at the Berkeley astronomy department in 1992 was that the discovery of the Kuiper belt gave Pluto a context. It took me and most other astronomers a few more years to realize that Pluto is neither lonely nor an oddball, but rather part of this vast new population called the Kuiper belt. Just as the explosion of asteroid discoveries 150 years earlier had forced astronomers to reconsider the status of Ceres, Pallas, Juno, and Vesta and change them from full-fledged planets to simply the largest of the collection of asteroids, the new discovery of the Kuiper belt would certainly force astronomers to reconsider the status of Pluto. It was becoming more and more clear that if the asteroids were the schools of minnows swimming among the pod of whales, then Pluto and the Kuiper belt objects were simply a previously overlooked collection of sardines swimming in a faraway sea. If Ceres was to be thought of as just the largest of the vast collection of asteroids and thus not a planet, why should Pluto not suffer the same fate? What, after all, was a planet?

Chapter Three

THE MOON IS MY NEMESIS

When I first started looking for planets, I lived in a little cabin in the mountains above Pasadena. I have a feeling I was the only professor at Caltech at the time who lacked indoor plumbing and instead used an outhouse on a daily (and nightly) basis. I worked long hours, and it was almost always dark, often past midnight, when I made my way back into the mountains to go home for the night. To get to my cabin, I had to drive up the windy mountain road into the forest, past the national forest parking lot, and down to the end of a dirt road, and finally walk along a poorly maintained trail by the side of a seasonal creek. For some time after I first moved in, I tried to remember to bring a flashlight with me to light my way, but more often than not I forgot. On those nights, I had to navigate the trail by whatever light was available or, sometimes, by no light whatsoever.

The time it took to get from the top of the trail to the bottom, where my cabin was, depended almost entirely on the phase of the moon. When the moon was full, it felt almost like walking in daylight, and I practically skipped down the trail. The darker quarter moon slowed me a bit, but my mind seemed to be able to reconstruct my surroundings from the few glints and outlines that the weak moonlight revealed. I could almost walk the trail with my eyes closed. I had memorized the positions of nearly all the rocks that stuck up and all the trees and branches that hung down. I knew where to avoid the right side of the trail so that I wouldn't brush against the poison oak bush. I knew where to hug the left side of the trail so that I wouldn't fall off the twenty-foot embankment called "refrigerator hill," named after a legendary incident when some previous inhabitants of the same cabin had hauled a refrigerator most of the way down the trail before losing it over the edge and into the creek at that very spot.

I had *almost* memorized the trail, but every twenty-nine days I was reminded that there is quite a big difference between memorization and near memorization.

Every twenty-nine days the moon became new and entirely disappeared from the sky, and I was almost lost. If by luck there were clouds that night, I might be able to get enough illumination from the reflected lights of Los Angeles, just a few miles away, to help me on my way. But on days with no moon and no clouds and only the stars and planets to light the way, I would shuffle slowly down the trail knowing that over here—somewhere—was a rock that stuck out—there!—and over here I had to reach out to feel a branch—here! It was a good thing that my skin does not react strongly to the touch of poison oak.

These days I live in a more normal suburban setting and drive my car right up to my house. I even have indoor plumbing. The moon has almost no direct effect on my day-to-day life, but still, I consciously track its phases and its location in the sky and try to show my daughter every month when it comes around full. All of this, though, is just because I like the moon and find its motions and shapes fascinating. If I get busy, I can go for weeks without really noticing where it is in the sky. Back when I lived in the cabin, though, the moon mattered, and I couldn't help but feel its monthly absences and the dark skies and my own slow shuffling down the trail.

Contrary to the way it might sound, however, back then the moon was not my friend. The two-and-a-half-year-old daughter of one of my best friends—a girl who would, a few years later, be the flower girl at my wedding—would say, when asked about the bright object nearly full in the night sky: "That's the moon. The moon is Mike's nemesis." And indeed, the moon was my nemesis, because I was looking for planets. Astronomers build telescopes in the most remote places they can find—the mountains of Chile, volcanoes in Hawaii, the plains of Antarctica, even in outer space—partially in the hope of escaping the city glare that increasingly permeates the skies. For all that effort, though, we can't hide from the brightest light that illuminates the night skies and washes out the faintest stars: the full moon.

As a new graduate student in astronomy at Berkeley, I had never previously considered the moon to be an obstacle. It was still the world that people had walked on early in my childhood, the scene that I'd drawn picture books of, the thing I'd tried to reproduce in my muddy, rock-splattered backyard; it was not a menace to be avoided. But I soon learned the lingo: Nights when the moon was full or nearly full were called "bright time" and were to be avoided by serious astronomers looking for faint objects in the sky. Times when the quarter moon was out for half of the night were "gray time." But the coveted nights were those when the moon was new and didn't disturb the dark sky at all. Only on those nights—"dark time"—do astronomers have a hope of detecting the very faintest blips of light that their telescopes can possibly see. I was now looking for planets, and a distant planet would indeed be a faint blip of light that the full moon would thoroughly overwhelm. So the moon became my nemesis.

I had started looking for planets by accident. In 1997 I began working as an assistant professor at Caltech, and I realized that I didn't really know what I was doing. Caltech is one of the best places in the world to be an astronomer. The university owns an inordinate number of the largest and best telescopes in the world, so Caltech astronomers are always expected to be—and often are—the leaders in their fields. When I started at Caltech, at the age of thirty-two, I suddenly had access to all of these premier telescopes, and I was told, essentially: Go forth! Use these telescopes to lead your field to new great things!

I had spent most of the six years of my Ph.D. studying Jupiter and its volcanic moon, but it was time to start something new, and here was my chance. Go forth! I thought. Okay. But where? Sure, I knew how to use the telescopes and the instruments and how to point them at the region of the sky in which I was interested, and I knew how to collect and analyze the data. But figuring out where to point the telescopes in the first place and why you're doing it is much harder. I was thoroughly overwhelmed. But I would not last long as an assistant professor if I didn't discover something big soon. I took out the list of all of the telescopes with all their capabilities and thought and stared.

It had been five years since that afternoon when Jane Luu had first told me about the Kuiper belt, and by this point almost a hundred small bodies were known in distant orbits beyond Neptune. It was becoming increasingly clear that the study of these very distant, very faint objects was going to be a major new field in astronomy. Big telescopes are particularly good for studying very distant, very faint objects, and I suddenly had big telescopes at my disposal. Go forth! I thought.

I didn't quite boldly go forth; instead I took a tiny step. I set out to test one of the hypotheses that was floating around in the scientific community at that time: that the objects in the Kuiper belt have mottled surfaces owing to the effects of craters formed by giant impacts, just like those that I could see on the moon. Proving or disproving this hypothesis would not be considered by anyone to be a major scientific advancement, but it was a start, and I needed a start. To test the hypothesis, I was going to spend three nights at the 200-inch Hale Telescope carefully studying a few objects out in the Kuiper belt to see if their surfaces were indeed mottled. The three nights I was scheduled to be at the telescope happened to fall over Thanksgiving, a fate that often befalls the youngest astronomer on the block. But the three nights were dark time. There would be no moon to disturb my view.

A day before Thanksgiving, I took the three-hour drive south from Pasadena, across the farmland (now housing developments) of the Chino Hills, through the dusty Pala reservation (now a multistory casino), and into the forested road (now a road through burned stumps) leading to Mount Palomar. The drive gives ample opportunity to stare at the sky and fret about occasional clouds and potential bad weather. This day there were no occasional clouds or potential bad weather: There was total cloud cover and continuous bad weather. The forecast was bleak. Astronomy is not always about bad weather at telescopes, but when you are young and eager for discovery or even just a few small steps, the nights of bad weather are the ones that seem to stick most in the mind.

The fog settled in thickly around the mountain as I arrived at the top and drove to the ornate old two-story heavy stucco dining and sleeping area known as the Monastery (which was an appropriate image for the earlier days of astronomy, when women were not allowed to stay). I went to the telescope to set up the instruments for the night of work; I spent hours in the windowless dome testing and calibrating and double-checking all of the settings I was going to use. As I finally stepped outside to walk to dinner, a light snow began to fall. After dinner the snow stopped, but a dense fog remained for the night. I stayed awake the whole time, hoping that somehow the fog would lift and I could start working. But it never did. I finally left the telescope to head back to the Monastery as the sun was rising and turning the fog from thick and black to thick and vaguely gray. At the Monastery, I closed the blackout curtains in my tiny room and slept until 2:00 p.m.

Opening the blackout curtains, I was greeted with more fog and now a heavy covering of wet snow. I was informed that the snow meant that there was no chance the telescope would be working that night; the dome enclosing it was frozen shut and would require direct sunlight to get unstuck. The snow also meant that the roads up and down the mountain were impassable in my two-wheel-drive truck. Instead of a quick meal before sunset with the other astronomers so that we could all run to our different telescopes when darkness arrived, we were all stuck at the Monastery for Thanksgiving. There was no television and no Internet connection, so after dinner, the other astronomers and I built a fire and caught up on our scientific reading. I was still scouring everything I could find to help me come up with ideas of what I might do. Every time I had a thought, I would ask the others around the fireplace questions about the local telescopes and how I could use them to help with this problem or that.

"How well does the infrared camera at the Hale Telescope

work?" Very well, was the answer. A general conversation would follow. We would all drift back to our reading.

"Is there a long-slit mode for the echelle spectrograph?" I would pipe out. No, was the answer, but we all speculated about how a quick modification would make one possible.

"Does anyone know anything about the new thermal imager that is coming next year?" Yes, indeed.

During the course of the evening, I covered, I thought, every combination of telescope and camera and spectrograph and instrument that was available at Palomar.

Eventually one of the other astronomers asked: "Have you ever thought about the 48-inch Schmidt Telescope?"

No. I hadn't. In fact, I only vaguely knew where it was. Down one of those side roads I never drove down? That little dome over by the water tower, maybe?

I did know, though, that when astronomers were building the huge 200-inch Hale Telescope more than fifty years ago, they realized that having the biggest telescope in the world didn't do you much good if you didn't know where to point it (a dilemma with which I am quite familiar). They decided that they needed to make a detailed atlas of the entire sky—a road map for the big telescope. So they built a smaller telescope, then known simply as the 48-inch Schmidt (after the size of the mirror and the general type of telescope), just down the road. The 48-inch Schmidt took pictures of the sky night after night until finally—for the first time in history—every patch had been photographed. The resulting maps of the skies—the Palomar Observatory Sky Survey—are famous throughout the astronomical world. At one time, all astronomy libraries had a wall full of cabinets containing fourteen-inch-square prints that together made up the complete Palomar Observatory Sky Survey. Each print, when pulled out of its special protective envelope, shows an area of the sky that looks about as big as your fully outstretched hand held at arm's length. It takes 1,200 of those prints to cover the whole sky, from Polaris, the North Star, all the way down to the Southern Cross.

As a graduate student, I had been instructed in the arcane mysteries of the correct use of the Palomar Observatory Sky Survey, which was simply called POSS by the cognoscenti. First, you go to the astronomy library and open the big cabinets; then, based on the sky coordinates of where you want to be looking, either you find the library ladder and climb to the top (if you're looking in the far north), or you sit on the floor (for the farthest southern objects), or, if you are fortunate enough to be looking for something directly overhead, you can stand comfortably and look straight ahead. With luck, you will find that the prints are stacked in the order they are supposed to be in, from pictures of the sky farthest to the east to pictures of the sky farthest west. If you're unlucky, the one picture you're looking for will be the only one out of place, and your search might take an hour. When you find the picture you want, you pull it out, set it on the large library table, put your face down close to the picture to see the millions and millions of stars and galaxies, and use the jeweler's loupe to find the precise area of the sky you're looking for. Finally, you pull out a custom-built Polaroid camera from its case, point it at the spot you've identified, and take an instant picture of a postcard-sized section of the sky survey print. That Polaroid print is now your personal road map.

For decades, astronomers carried these Polaroids with them to telescopes all around the world. When you commanded your large telescope to point to the spot in space in which you were interested and you looked at the TV screen, you were usually greeted by a fairly unremarkable field of stars. The Polaroid pictures were the only way to know that the unremarkable field you were looking at was the one in which you could find the galaxy or the nebula or the neutron star you were looking for. In the control room of any telescope at any night of the year, you could find an astronomer or a group of astronomers holding a Polaroid print and staring at the TV screen. Often the actual image of the sky from the telescope was flipped or upside down and no one could ever remember which particular way this combination of instrument and telescope flipped images, so there would always be a time in the night when three or four astronomers would be squinting at a little screen full of stars, holding a little Polaroid picture full of stars, and turning the picture sideways and upside down until someone exclaimed, "Ah ha! This star is here, and that little triangle of stars is here and we're in just the right place." These days the technique is mostly simpler—the Palomar Observatory Sky Survey pictures are all quickly available over the Internet, and the cabinets full of prints are gathering dust; but because you can't take the computer screen and turn it sideways or flip it over, the little group of three or four astronomers is now more often than not standing with their heads cocked in all possible combinations of directions until the lucky one exclaims, "Ah ha!" and then all heads immediately tilt to that direction.

Though the 48-inch Palomar Schmidt was famous to astronomers the world over, I had not even considered it worthy of thought, for one good reason: The telescope still used relatively primitive photographic technology to take pictures. Astronomers a generation before me all learned photographic astronomy: how to load film in the dark, how to ride in a tiny cage suspended at the top of the telescope, how to carefully move the telescope through the sky, how to develop and print. My generation was the first entirely digital generation of astronomers. All telescopes today have digital cameras that use, in only slightly fancier form, the same technology used in everyone's handheld digital cameras around the world. The change in astronomy is as dramatic as it has been in photography. The ease and speed with which images can be obtained and examined and manipulated and shared has transformed the way that astronomy is done today. So when I overlooked the 48-inch Schmidt Telescope, it was mostly because I considered it a relic from the days of prehistoric astronomy.

But on that snowy, foggy Thanksgiving night at Palomar, I decided that visiting this relic to see how ancient astronomy used to be performed would be an entertaining way to spend a few of the nighttime hours. After making sure I knew exactly which way to go, I walked down the dark, snowy road through the piney woods, past the largest telescope, down a road I had never taken, to where the 48-inch Schmidt resided. Someone was inside, tidying up in the cramped control room that sat underneath the telescope. I introduced myself and met Jean Mueller. She was tidying, in lieu of her usual nighttime job, which was to use the 48-inch Schmidt to once again make a new map of the full sky to compare to the first.

Using the 48-inch Schmidt? It was a fossil. Why would anyone still want to use it and its messy and cumbersome photographic plates? The answer is relatively simple. Even though astronomy has progressed greatly since the days of photographic plates, and even though digital cameras make astronomers' lives incomparably easier and better, one thing had gotten worse. A Schmidt telescope is designed to look at a huge swath of sky at once. Every time a fourteen-inch-square photographic plate—literally just a piece of glass with photographic emulsion painted on one side of it—is placed at the back of the telescope and exposed to the night sky, an enormous piece of the sky is photographed. Digital cameras on telescopes, in contrast, are much better at seeing faint detail but much worse at seeing large swaths of sky. A typical telescope equipped with a digital camera could, at the time, only see an area of the sky more than one thousand times smaller. The obvious solution would simply be to build a bigger digital camera, but to see as much sky as you could see with the photographic plate you would need a five-hundred-megapixel digital camera. Even today that is a daunting number. At the time, when only high-end photographers had a single megapixel to their name, if you wanted to make a map of the sky, just as the 48-inch Schmidt had done in the 1950s, it made much more sense to accept the hardships of the photographic plate for its unparalleled ability to sweep up the night sky at a fast pace.

Jean explained this latest survey and described how the photographic plates were taken and developed. She talked about how she had come to be working there at Palomar after a few years at another observatory. She then wistfully told me that the days of the 48-inch Schmidt were almost over. This second Palomar Observatory Sky Survey was almost complete, and she didn't anticipate that anyone else would be using the telescope and its photographic plates after that. All of the fall sky had already been photographed, and no one planned on using the telescope at all during the fall season the following year.

All major telescopes around the world are scheduled to be used every single night of the year, with the occasional exception being Christmas, though I've worked at telescopes on Christmas Day plenty of times myself. I find the idea of a telescope not being used almost viscerally painful. It's bad enough when the reason is technical or simply weather related, but when a telescope is not being used for simple lack of interest it feels worse.

Yes, the photographic technology was old and clunky, but clearly the 48-inch Schmidt was one of the best telescopes in the world, at least for imaging wide areas of the sky.

Wide areas of the sky! This was just what I needed! The study of the Kuiper belt, still in its youth, was hampered by the fact that astronomers had been searching for objects in the Kuiper belt with digital cameras that covered only small areas of the sky at once. They were successfully finding objects, but the objects were all small and faint. Imagine being interested in exploring the inhabitants of the ocean but all you have is a small handheld net. If you dip your net in the sea many times, you will certainly find a vast collection of microbes and krill, but you will never know that there are dolphins and sharks and even the occasional whale. In contrast, the photographic plates from the 48-inch Schmidt were not nearly as sensitive as the digital cameras that other astronomers had used—the net was so large that the krill and the microbes would fall right through—but we had a net big enough that we could cover the whole ocean. The big fish would have nowhere to hide.

I thought about the biggest fish.

I had already been thinking by this time that Pluto might not be a solitary planet out there in the Kuiper belt; there might be others still to be found. And using the Schmidt was clearly the way to find them. There was a major problem, though. The last time I had touched real film was when I was in third grade and my father and I had built a little darkroom and developed our pictures from the pinhole cameras we made. There was no way I could carry off this project. I gingerly inquired as to what Jean was doing next fall, when the telescope was to be idle. She didn't know. She and her coworker would presumably be assigned other tasks around the observatory during that nonworking season. And what if someone else was interested in using the telescope? I asked. Her face lit up as she quickly exclaimed, "I'm sure everyone would be thrilled—we would love to have new projects on the telescope."

Then she asked: "Do you think we might find a planet?"

. . .

And so I came to be looking for planets. A year later I got to know Jean and her coworker Kevin Rykoski extremely well, as every night, except for bright time, when my nemesis interfered, I called in to talk about what section of the sky to photograph that night. Every night, in all possible permutations, we discussed the position and phase of the moon, the possibility of clouds or fog, and the success or failure of the pictures from the night before. Everywhere I went I carried my hardbound notebook containing maps and calendars and records of everything that we had done to date. Every night, no matter what time zone I was in or continent I was on, I called in to the 48-inch Schmidt precisely thirty minutes before the sun set (the time of which was recorded for every night in my black notebook). I remember making the call from a pay phone on a busy evening street in Berkeley, early in the morning from a hotel in northern Italy, well past dark from my mother's house in Alabama, but most of all from that little cabin in the woods.

I had meticulously worked out the procedure. Every month we would cover fifteen separate fields, or an area that covered a little over 1 percent of the full sky. While that doesn't sound like much, in just a single month we were going to have covered more sky than all other astronomers searching for Kuiper belt objects had covered in the preceding five years. On a typical night, we would try to cover three or four fields. To do so, Kevin or Jean would walk from the dimly lit control room crammed with computer equipment and go up a winding set of stairs to the floor of the telescope dome. Once inside, all of the lights would be put out as they would unpack one of the photographic plates from where it had been stored in a light-tight box. From my pinhole camera days I remembered that film was developed in red light, which doesn't affect it. But these photographic plates were designed to be *especially* sensitive to red light, as objects in the Kuiper belt tend to be on the red side. All of the work on the plates, then, had to be done with no lights whatsoever. When the plate was unpacked, it would be walked to the telescope and inserted into the base. Only then was the shutter of the telescope opened and the light from the sky allowed to beam onto the plate. Thirty minutes later, someone would again walk to the top of the stairs in the dark, take the plate out of its holder, and then walk to the other side of the dark dome floor and place the plate into a miniature manual elevator and drop it down to the other person who was waiting in the darkroom below. The person on the dome floor would get a new plate and begin looking at a new patch of sky, while the person in the darkroom washed the plate in a succession of developer and fixer fluids until, about the time that the plate was finished, a new plate would appear in the miniature elevator. In the morning, before going to sleep, Jean and Kevin would look at the crop of pictures from the night. Some would be smeared or have defective photographic emulsion and have to be rejected, but the good ones got labeled, put into the cabinet, and filed on my list. The next night we would review what had happened the night before, discuss the weather forecast, curse the encroaching moon, and start over again.

I found this exhausting, and I was the only one of the three of us actually sleeping at night.

The goal was to get three good images of each of the fifteen fields during the course of the month. Ideally they would be taken three nights in a row. My job was to examine each of the images and, as astronomers had been doing for two hundred years, look for the things that move.

Kevin and Jean must have been happy that the moon existed, since bright time was the only time that they got a few days off. But I was no fan of the moon. I became increasingly agitated as the month progressed from gray to dark to gray again and finally the bright approached. Invariably as the month was coming to an end we would be behind schedule owing to problems with the weather or problems with the photographic plates. I would count ahead the number of nights left before bright time commenced and almost always find that everything had to go perfectly or we would lose one of our fields. And every lost field meant that any planets out there in the sky suddenly had a huge place to hide. Our net would have holes. Near the end of the month, Jean and Kevin would invariably work overtime. I could do nothing except sit in Pasadena, stare at the moon, and fret.

Somehow, we managed. In two years of surveying the sky with the 48-inch Schmidt Telescope, we actually managed to get every image of every field we wanted except for one. We mostly beat the moon. Final score: 48-inch Schmidt, 239 fields; moon, only one field. Those 239 fields we had covered were only about 15 percent of the whole sky, but it was, we thought at the time, the right 15 percent. The moon and planets are all strewn across the sky in a giant ring encircling the sun, and we had looked at that ring—as well as a good bit above and below—for a period of about four months, or one-third of the whole ring. So while we had looked at a relatively small fraction of the sky, it was much of the interesting sky, and it was enormous compared to what had been previously examined. We hadn't taken our net through the entire ocean, but thought we knew one of the whales' major swimming grounds, and we had trawled it all.

Looking at vastly more sky than anyone else had ever looked at for large objects out in the Kuiper belt was so immensely exciting that I could hardly contain myself. I *knew* that there would be big discoveries, and having new pictures come in night after night after night with only a break for the full moon kept everything at a constant peak. I talked to my friends about new planets. I thought about names for new planets. I gave lectures about the possibility of new planets. I did everything I could, except find new planets.

Of course, I did more than talk on the phone and make sure that the pictures got taken. After each set of pictures was exposed, the photographic plates would be put into large wooden crates and shipped from the mountaintop down to my office in Pasadena, where my work would begin. I needed to turn those crates full of plates into discoveries of planets.

Seventy years earlier, Clyde Tombaugh found Pluto by doing almost exactly the same thing that I was currently doing, except that he did all of the work himself. He would stay up all night exposing the photographic plates to the sky, and then in the daytime he would look for things that moved. To look, he would take a pair of photographic plates that showed the same region of the sky and then load them into a specially constructed apparatus the size of a large suitcase, called a "blink comparator." Inside the blink comparator, a light would shine through one of the plates and project an image toward the top, as if the photographic plate were a giant slide. On the outside, Tombaugh could look into the comparator with an eyepiece and have one of the slides projected into his view. The special part, though, was a little mirror inside that could quickly flip back and forth so that Tombaugh could look at one of the photographic plates and then the other in as quick a succession as he wished. All of the stars in the sky, all of the galaxies, all of the nebulae, would appear the same on each of the two plates, but anything that moved or changed or suddenly appeared would jump out as the two photographs were blinked.

Palomar Observatory had had a similar apparatus to Tombaugh's in its early years, but it had been disassembled a couple of decades earlier. But even if such a thing *did* still exist, it would have done me no good. Because the telescope that I was using was so much more powerful than the one Tombaugh had used to find Pluto, each of my images showed one hundred times more stars, and thus would have taken one hundred times longer to have gone over by eye. Early on in the project, I calculated that to look at every star on every photographic plate by eye would have taken me forty straight years of staring into the blink comparator and slowly watching pictures of the sky go by.

Not wanting to wait forty years, and it being 1998 instead of 1930, I put the computers to work instead. First, we needed to scan the photographic plates to get them into digital form, and then the computer could do the rest. The scanning was quickly done on a big machine that already existed. Getting the computer to do the rest, though, took longer. There is no software package that looks for planets. I would have to write it myself. Though I knew nothing about emulsion and developer and fixer, this I could do. This I was good at. I had been writing little computer programs to analyze and predict and follow the stars and moons and planets in the night sky since high school. This would be the first program that actually mattered.

I spent most of that year slouched in front of a computer screen in my office, testing, scowling, starting over, typing furiously, and pondering. For someone looking for planets, I spent an awful lot of my time looking at computer code and numeric outputs instead. My nights were spent not outside staring at the sky but inside staring at numbers and computer programs and doing every test conceivable. I needed to make sure the software wasn't going to make any mistakes. I wanted to make sure that I didn't do anything stupid that made me miss planets that were right in front of me.

I made the computer begin by looking at a triplet of scanned photographic plates. It examined each of the little blips of light on each of the three images taken over the three nights. All of the stars in the sky, all of the galaxies, all of the nebulae, had the same coordinates on each of the three photographic plates, so the computer quickly identified them as not moving and tossed them aside. Sometimes, though, something appeared at a spot in one image where the other two images showed only blank sky. The computer took note. It could be many things. Sometimes stars in the sky get brighter and suddenly show up where they weren't seen before. Sometimes satellites in orbit around the earth give a sudden glint that looks like a star. Sometimes dust blowing around at night sifts through the open shutter of the telescope and settles down on the photographic plate, disturbing the precarious emulsion and making something that looks vaguely like a star. But sometimes something appears where it has never appeared before because it is slowly wandering across the sky and that single picture happened to catch it momentarily in one spot. In that case, an image the next night would find it again, only a little displaced from the previous night. I used the third picture as a final check. When the computer found a third object that looked as though it could be connected to the first two, it put that object on a list of potential new wanderers and moved on to the next spot in the sky. All of this takes, of course, about a millisecond. To process our two years' worth of images took under two hours.

So after Kevin and Jean had spent all of those nights loading and developing plates, and I had spent a year programming the computer, and the computer had spent two hours processing all of the final data, I finally had a list of all of the potential new planets to look at. I had been sustained throughout this time by the thought of this moment. I was going to find a planet, and the solar system would never again be the same. When I first opened up the list on the computer screen and started scrolling down, I must have gasped. The list was 8,761 candidates long.

I knew that the computer would be overzealous in identifying potential planets; in fact, I had written the program to make *sure* that the computer was overzealous. I had decided early on that I would make the computer find everything even remotely possible, and I would look at each thing the computer picked out by eye to double-check it. But 8,761 objects to check by eye was going to take a long time.

I slowly began to go through the list. I would press a button on my computer, and on my screen three pictures would appear of the three nights of the same small region of the sky, with little arrows showing where the potential planet lay. I saw an amazing number of small glitches that had fooled the computer. Scratches on the photographic plates, of which there were many, would cause a star to disappear one night and so appear as if it were new the next. Anyone looking at the pictures could see that it was just a scratch, but to the computer it appeared as dark sky. Sometimes the light from a particularly bright star would get reflected around in the telescope perhaps dozens of times and give tiny apparent glints all across the sky. By eye, you would notice all of the glints and you'd see the proximity of the bright star, and you would quickly say, "Ah, that's just a bright star making glints," but to the computer it was a star never seen before.

The examination took months. On the computer screen, I had a "no," a "maybe," and a "Yes!" button that I chose from after examining each of the pictures. Had it been a mechanical button instead of a virtual one on the computer screen, I would have worn the "no" button through. The "maybe" button got a little bit of action, too. Sometimes I would look at three pictures and find no obvious problems with what the computer had done, but I still wasn't entirely convinced that what the computer had picked out was really there at all. The photographic emulsion was sometimes a little uneven, and the computer might have picked out a slightly brighter spot that really was just the sky. A tiny speck might appear that was possibly a faint star, but I was not quite convinced. In all of those not-entirely-sure cases, I would simply press "maybe."

"Yes!" was reserved for the no-questions-no-problems-definitely-really-there-moving-through-the-sky cases. Every day I would come in thinking that perhaps today was the day that I would finally push the "Yes!" button. Every day I would spend hours staring at the computer screen, pushing "no," and occasionally, very occasionally, "maybe." But the "Yes!" button remained untested. After going through the entire set of potential planets that the computer had picked out, I never once used the "Yes!" Final score: "no," 8,734; "maybe," 27; "Yes!" 0.

It was hard not to feel distressed. What if there really *were* no other planets out there? What if three years of photographing and computing and blinking came down to nothing at all? What if the big project designed to make my splash as a young professor at Caltech disappeared without a ripple? I had been telling everyone for three years now that I was looking for planets, that I was going to *find* planets. What if there were no planets?

I still had hope, though, in the twenty-seven maybes. I spent much of the fall of 2001 at Palomar Observatory trying to track them down. For a few dark nights every month, I would drive to the mountaintop, arriving early in the day to plan for the night and prepare the telescope, eating dinner before the sun was close to setting, packing up a bag full of truly awful snacks designed to keep me awake throughout the night, and then heading for the control room of the 60-inch telescope.

This telescope had a modern digital camera, which meant that it was quite sensitive but that it covered a tiny area of sky. The nights were carefully choreographed so that I could spend the most time looking at the expected locations of the twenty-seven maybes. Because a full year had passed, they had moved quite a ways, and it was impossible to know precisely where they might be, so I would spend hours scouring large parts of the sky, taking a picture, and coming back to the same spot an hour later and taking another picture. I didn't even bother writing a computer program for these; I would just look at the blinking images on my computer screen the second that they came down from the computer. All night, every night there, I would take a picture, move the telescope over, immediately start another picture, stare at the last picture while taking the current picture, and continue on until dawn. Then I would slowly and wearily walk the winding road the half mile back to the Monastery, often startling foxes or bobcats out for a dawn hunt. Around noon, I would wake up, have breakfast, and begin the day again.

During those first few months tracking down maybes, I felt excited when the sun went down.

Tonight is the night! I would think.

As the fall progressed, though, I was slowly becoming dejected.

I spent so much of my time at Palomar Observatory that fall that I didn't have to think twice when I got a request to give a talk at the observatory to a group associated with Caltech. I was going to be there the night before anyway, so I figured I might as well stay one more night to give the talk. On my calendar I just wrote "talk to some group." The group was to arrive by bus in the late afternoon, take a tour of the massive Hale Telescope, and eat dinner and hear my talk on the floor of the dome with the telescope perched overhead. It sounded fun. I like giving talks to groups like this.

The afternoon the group was to arrive, I waited on the dark ground floor of the observatory until I heard a knock on the door. As I opened it, I was blinded by the afternoon sun. When my eyes adjusted again, I finally saw the organizer of the tour walk in.

"Hi, I'm Diane Binney," she said.

She was well dressed, poised, glamorous, outgoing, radiant. She was everything that you don't stereotypically expect to find in someone from Caltech (including, in particular, me). I quickly introduced myself, and I thought: Who *is* this person?

Diane Binney was the well-loved director of a group whose members attended tours and special talks and traveled to exotic locations, all associated with Caltech and its research. Diane had arranged this trip to Palomar Observatory and had invited me to speak, and, as I learned much later, everyone except for me on the Caltech campus seemed to know precisely who she was and had known for years. I had perhaps been staring into my computer screen too much to have ever looked up and noticed.

I admit that I did not give the people on the tour the full attention that they deserved. I admit to spending more time telling Diane about the telescope and the dome and astronomy than I did everyone else. But I must have given an all-right tour—at least to her—because at some point while walking on a catwalk high above the ground on the outside of the observatory, she said, "Hey, do you ever use the telescopes in Hawaii?"

I do.

"Would you be interested in coming next spring on a travel program where we take people to the volcanoes and then up to the telescopes? Would you be able to talk about the telescopes and give tours?"

Not checking my calendar, I simply said, "Absolutely."

Dinner soon began. I spoke for an hour and showed pictures of the sky, pictures of telescopes, and graphs of what was to be found out at the edge of the solar system. But mostly I talked about planets. I told the group that there *had* to be planets out there and that I was going to find them. Even as I said it, though, all I could think of was that I was halfway through my "maybe" list, and still I had found nothing. I could do the song and the dance and put on the excited face, but it was becoming possible that all of my searching would come to nothing.

When the talk was over, the group got on their bus and left. I walked over to the little cottage where Kevin Rykoski lived. I had talked to Kevin and Jean Mueller on the phone every night discussing where to point the telescope, but now I finally had a chance to go sit on Kevin's sofa and drink a beer. He had been at my talk earlier and had helped with the tour.

Over time, my conversations with Kevin and Jean every night while taking the photographic plates had progressed from simple efficient talk about the sky and the weather to a more general extended chat. Jean would talk about her plans for a dream house on the river, while Kevin told stories about his teenage daughter or described how he would drive directly to the beach on the last morning before bright time started and sleep all day long. Kevin and Jean had also had inadvertent front-row seats to the demise of my long-term relationship from my days in Berkeley and my subsequent retreat from the cabin in the woods that my girlfriend and I had shared, to the death of my father, to the start and end of a new relationship or two; so, as I sat on Kevin's sofa for the first time, our conversation naturally steered to the personal.

All Kevin wanted to talk about was Diane Binney and why she kept talking to me. I told Kevin about the Hawaii trip and that we were talking logistics. He thought that sounded like an exciting first date. I insisted that it sounded like work, because that was all it was.

Kevin wouldn't let up. "Yeah, but she was paying you a lot of attention."

"She runs trips for people; it's her job to be nice. I'm sure that all of the guys at Caltech that she has to work with get the wrong impression and make idiots of themselves. I'm not going to do anything stupid."

Six months later, I was in Hawaii with Diane and twenty or thirty people in her group. The group spent an enjoyable week on the lava, at the telescopes, at the beach, learning about geology and listening to me lecture about astronomy. The last night, when she was done with the trip and could finally relax, the two of us found ourselves down on the beach alone sometime after midnight. I pointed out the Southern Cross, just barely visible at the right time of year from Hawaii, and I showed her the paths of the planets and how she could pick out Saturn just setting into the ocean. I told her what it was really like to use the telescopes, and she talked of her nieces back in California. Saturn sank into the Pacific, and we finally walked back to our rooms. I was quite proud of myself for not having done anything stupid.

When we got back to Caltech the following week, I found myself accidentally walking past Diane's office a few times a day and accidentally running into her and stopping to talk. Every time I did, she was very nice, and I had to remind myself that, truly, it was her job to be nice and to appear happy to see me and that being stupid was the worst thing to do. On accidentally running into her in the early afternoon one pleasant spring workday, I asked if she needed a cup of coffee. She did. We walked down the street, drank coffee, and talked for three hours. Certainly, it was part of her job to be nice to me and cultivate me as a good resource. But it occurred to me that, even accounting for all of that, there was no reason for her to spend three hours in the middle of an afternoon with me when we both had many other things to do. It suddenly occurred to me that, in fact, I had been stupid all along.

Later that summer, when Diane and I went on another trip together, I did no astronomy lecturing, and she brought no group. Instead, the two of us spent a week in a little cabin on a tiny island north of Vancouver. It was the least stupid thing I had ever done.

Sometime during this period it became clear that all of my searching for planets was going to come to nothing, that all of the maybes were turning into definitely nots. Three years of intensive effort to find a planet was leading to the conclusion that nothing was out there to be found. I don't actually remember when I finally closed the black hardbound notebook for the last time. I don't know when I really admitted to myself that there was nothing there. In fact, I don't remember much at all about planets and searching for them around this period. All of the irritation of the quickly dismissed nos and all of the frustration of the maybes that I had spent nights and nights at the telescope trying to track down suddenly seemed much less interesting than trying to figure out the next trip I would be asking Diane to take with me, to which she would inevitably say "YES!"

Chapter Four

THE SECOND-BEST THING

In June of 2002, Chad Trujillo walked into my office and announced, "We just found something bigger than Pluto." It turned out to be the second-best thing that happened to me that week, and it wasn't even true.

Chad had recently moved from Hawaii to California to work with me on a brand-new project: using the 48-inch Schmidt Telescope at Palomar Observatory to find planets. Yes, the project sounds familiar, and yes, I had spent three years using exactly the same telescope to find exactly the same planets, and I had thoroughly failed to find anything. Yes, I had even been strongly advised by people who were concerned with my career and who had influence over whether I, a young assistant professor, stayed at Caltech or not, to quit this planet searching altogether and do something more respectable. But really, how could I just stop? Sure, we had looked at a lot of the sky—more than anyone else since Clyde Tombaugh had discovered Pluto seventy years earlier—but we hadn't looked at the *whole* sky. So how do you know that you've done enough? If there is only one or two or even just three new planets out there waiting to be found, what are the chances that you just didn't look in quite the right spot? And how could you really convince yourself that there was nothing there unless you looked in every single corner where things might be hiding? Maybe the whales really had just slipped through the net.

In the two years after I finally declared my initial search unsuccessful, every once in a while I would get a phone call or an e-mail from a friend who remembered that I had spent a long time talking about searching for planets, and the friend would invariably say something like "Hey, I just read in the newspaper that someone found a new planet, did you hear about it?" My breathing would stop while my pulse doubled as I tried to casually use my now-shaky fingers to quickly search my computer for the news of the day. "Oh, no, I haven't heard, so, really, probably it's nothing. Nothing at all." At least I hoped. After all these years, the idea that someday someone would call me up and casually tell me that someone else had found a planet that I had missed still haunted me. Each time it happened, I would search the news and find, to the sudden restarting of my breath and calming of my pulse, that yes, indeed, a new planet had been found, but it was not a tenth planet orbiting around our sun, it was a planet orbiting a distant star far removed from our solar system. And I could then quickly tell the person how exciting it was that all of these new planets were being discovered around other stars and about how much we were learning, and how, oh, no, this was not the sort of planet I had been looking for at all. No one else was looking for planets out at the edge of our own solar system—at least that's what I thought. What I hoped.

Though my first search had come to nothing scientifically, planets were still never far from my mind. I still wanted to find one. I just needed a new way to do it.

Less than a year after the first failure, I was back to work on the sky, and this time I was determined to do the job right. It was 2001, and though perhaps Arthur C. Clarke's predictions of space tourism and obelisks on satellites of Jupiter had not come true, it was finally time to get rid of the hundred-year-old technology of the photographic plates. To some, it was a sad day when the photographic plate handling system at the 48-inch telescope was dismantled, though anyone who had ever had to work in the absolute darkness of the nighttime dome, moving plates from their holders to the telescope to the darkroom, could not have felt too bad to see it all go. The darkroom was turned into a storage room. The walls of the plate-handling room right next to the telescope were torn down to make room inside the dome. The mini elevator, which Jean had used innumerable times to transfer an exposed plate down to Kevin, who had been waiting in the darkroom, was permanently sealed. All of this was to make way for the new incarnation of the telescope: a modern, digital-camera-equipped, computer-controlled, remotely piloted sky-searching machine.

The difference between the digital camera and the old photographic plates was extreme. With the plates, you would walk upstairs, load the photographic plate, open the enormous shutter on the camera, and expose the film to the sky for about twenty minutes. It would take about ten minutes to unload the exposed plate and load in a new one and start all over again. In contrast, with the digital camera, you never had to walk up the stairs—indeed, you needn't even be awake! The computer opened the shutter, exposing the digital camera for about sixty seconds, and sixty seconds later you could be looking somewhere else in the sky. It took two minutes for the computer to do what it had taken forty minutes for Jean or Kevin to do earlier.

The digital camera was small, though, compared to the photographic plates, and it covered only about one-twelfth as much of the sky (an area equivalent to about three full moons)—but since it was twenty times as fast, we were still ahead. Even better, in the 60 seconds that the digital camera was exposed to the sky it was able to see stars and moons and planets that were two or three times fainter than the faintest things we had seen on the photographic plates. I had been worried for the past several years that the things for which we were looking lurked just beyond the limits of what we could see. I had stared at many of those photographic plates that Kevin and Jean had taken, and had wondered what we were just barely missing.

But now we were in business. We could run almost every clear night of the year without worrying about overworking anyone other than the computer. We could see fainter things. We could cover more sky. In the first four months, we planned to redo the whole region of the sky that had taken us three years to complete earlier. And then we were going to keep going. Surely, all of this would lead us to the planet that was still out there waiting to be found. I was certain we would find it quickly.

I was so certain that we were going to be finding things immediately that I decided I needed help. I recruited Chad Trujillo, who was just finishing his doctoral thesis—on, conveniently, finding objects in the Kuiper belt—at the University of Hawaii. I wasn't sure I was going to be able to convince him to come to Pasadena. He was so relaxed from his years in Hawaii that he seemed like someone more likely to live in a tree house than in a city, but having lived in the little cabin up in the woods in my earlier years in Pasadena, I knew where to find the most tree-houselike areas around. That, and the prospect of perhaps finding a planet or two, convinced him, and he moved to Pasadena and immediately set to work.

He knew what he was doing so well and was already so good at it that I essentially handed him the keys to the telescope and stepped out of his way. In a few short months, we'd finished looking at the area of the sky we had previously covered with photographic plates, and, to my great relief, there truly was nothing there to have been seen. Soon we were on to fresh sky. And somewhere in that fresh sky we made our first catch.

I'd love to write more about this very first discovery, about how Chad took pictures of the sky one night and then, while looking through them the next day, spotted a brand-new point of light slowly crawling across the images. I'd love to describe Chad's excitement as he stepped across the hall into my office and showed me that first discovery. Yes, it was somewhat small— larger things had been stumbled upon in the Kuiper belt already—but we now knew for certain that if we could find this relatively small chunk of ice so quickly, any planets hiding out there were going to be within our grasp. I should have felt quite vindicated after all those years of searching. Perhaps I even did. The only problem is, I don't remember any of it. When would it have happened? Probably November or December of 2001. Or was it January? Did Chad really come in and tell me? Did I then go across the hall and look at the pictures on his computer? It astounds me that I don't remember any of it. I could go back and look up the records of that first discovery and perhaps refresh my memory. Instead, I went and looked back at my calendar for that time period to try to remember what else was going on that fall and winter.

My calendar is filled with Diane: trips with her to Hawaii, to the San Juan Islands, to the Sierra Nevada, trips on which I didn't even check the phase of the moon before going. And when there weren't trips, there were dinners and coffees and lunches. A year earlier I had worked from 10:00 a.m. until 10:00 p.m. most days of the week; but in 2001–02, for the first time since I had become an astronomer, I was actually leaving work at an almost normal time. I wasn't even coming into work most weekends. I wonder if that first discovery came on a weekend when I wasn't there, or when Diane and I were off for a week snowbound in the mountains. I like to think so. I was confident in the future Chad was working on; I was confident in the future I was working on, too.

The distraction of the winter continued into the distractions of the spring. By now Chad was discovering new objects beyond Neptune at a steady rate. One or two were big and bright enough that despite my distracted state at the time, I can still remember their discoveries—though, admittedly, a little vaguely. By the beginning of the summer, Diane and I had planned a long-weekend escape to a small beach town in the Yucatán. Before leaving, I had had a long conversation with one of my Ph.D. students. For some of my students I play multiple roles, including scientific adviser, speaking coach, writing instructor, tool provider, caffeine enabler, and, sometimes, relationship counselor. This student was complaining about the fact that her boyfriend, to whom she was engaged, thought that buying an engagement ring when they were not to be married for several years made no sense. On the airplane ride down to the Yucatán, I told Diane the story, and added the opinion that not only did I agree with the boyfriend, but that I thought engagement rings were a silly waste of money anyway and wouldn't it be wiser to buy useful things? Like kayaks? Or bicycles? Didn't she agree? No, she didn't, actually.

I brought up the subject again at dinner our first night on the beach, and even once the next morning. Secretly, though, I had spent the previous month seeking out and buying the perfect engagement ring, and I had brought it with me. I had a plan. First, I would convince Diane that an engagement ring was the farthest thing from my mind; then I would plan a perfect seaside evening with dinner and a bottle of wine and spring it on her.

As I have since learned, I'm not very good at keeping burning secrets. Rather than waiting until the evening, sometime in the middle of our second day there, while sitting in a little hammock overlooking the water, relaxed and happy and talking about nothing in particular, I quickly went back to our room and came back out with the ring hidden in my pocket. I knelt. I proposed. I then went on for several verbal paragraphs about the symbolic importance of a ring as a combination of a public statement, something akin to earnest money, and a down payment on life. I then produced the ring. Diane was stunned silent. You could almost hear the machinery in her head reprocessing the last few days. Her first words, after a considerable pause, were: "You are such a shit." She continued processing the days, the conversations about rings, the fact that she'd thought I was hopeless. She wanted to know when I had gotten the ring (on the street right outside the hotel, perhaps?), who else knew (my mother, of course), why it fit perfectly (I had surreptitiously tried on her rings, and they all fit perfectly on my pinky, which I then had measured), how I had managed to pick one she liked so much (I modeled it somewhat after her grandmother's wedding band, to which I knew she was much attached). Finally I had to remind her that I had actually proposed and she had not, in fact, given me an answer. She looked up and said, "Yes!"

. . .

The week back on campus was filled with people congratulating me. In the middle of the week, the chair of my department stuck his head in my office and invited me out for a walk. He had known Diane years longer than I had, so I was expecting some sort of congratulations followed by a lecture on how to treat her. "Congratulations," he started, but surprised me instead with "you now have tenure."

"Oh," I remember saying. "Um. Thanks."

"Thanks? Usually people are a bit more excited to hear this news."

"Well, it's only the second-most-exciting news of the week."

Remarkably, receiving tenure at Caltech turned out to be only the third-most-exciting thing that happened that week. A day after my conversation with the chair of my department, Chad stuck his head in my office, looking no less relaxed than ever (perhaps having just come from surfing across town at Malibu), and said, "We just found something bigger than Pluto in the pictures from last night."

Bigger than Pluto! This one I remember. Not as calm as Chad, I rushed across the hallway to see the pictures on the computer screen. The night before, the telescope had photographed an anonymous patch of the sky near the Milky Way galaxy, and there amid the thousands of stars was one tiny dot slowly inching across the sky. Chad had determined how far away it was (almost 50 percent farther from us than Pluto), and from that and from the brightness had guessed that the object was probably bigger than Pluto itself. It was certainly the largest new thing that anyone had found in the solar system for more than seventy years. This is what we had been hoping for. Only a dozen people in human history had ever discovered anything bigger going around the sun. It was the second-best thing that had happened to me that week.

Chapter Five
AN ICY NAIL

There is a critical tension in science between the very human desire to announce discoveries immediately (both because you are excited about them and because you don't want to be scooped by someone else) and the very important need to carefully and systematically check and document your results. In some cases this documentation can take additional years of work. In the case of the new discovery that Chad had told me about, we were quite worried that someone else might stumble upon it in months or even weeks, so we put together a plan to try to learn everything we could about it in as short a time as possible, and we set ourselves a deadline of only four months to make an announcement, complete with a full scientific account of the discovery and anything else we could learn. For me, those months of trying not to tell anyone about our discovery was harder, even, than not telling Diane that I had had a ring in my pocket all along.

During this time, we decided that rather than repeating "the object that we just discovered," we should give it a temporary name. We settled on Object X. "X" was for Planet X, for unknown, and, perhaps, for tenth planet.

As scientists, we were eager to know everything about Object X, but the first question on our minds, the one that would put everything else into context, was: What sort of orbit did Object X have? Did it go in a circle around the sun like the planets, or did it have an elongated orbit like Pluto and the other objects in the Kuiper belt? To answer this, we would have to track the object through its orbit and learn where it went. This would take time and patience. Pluto, after all, takes 255 years to go around the sun. Object X, farther away, would take even longer. But time and patience were two things we could not afford. Fortunately, though, we didn't need to wait hundreds of years. We don't actually have to follow an object all the way around its orbit to know where it is going to go (a good thing, since we've still watched Pluto for only a little more than a quarter of its orbit). If something is moving under the influence of gravity alone, we need only to know precisely where the object is, precisely how fast it is going, and precisely what direction it is moving in to know where it was at all times in the past and where it will be in the future.

Even if you don't know how to work out the math yourself, your brain certainly does. Try this experiment. Stand in a field and have someone thirty feet away throw a ball in your direction (using a foam ball would be a good idea, as will become obvious). The second you see the throw, close your eyes and see if you can figure out where and when the ball is going to hit the ground. Chances are you'll do pretty well. Your brain is instinctively trained to quickly estimate the three key things—where, how fast, which direction—and predict where a projectile is going to go. But chances are you will not be precisely correct. The ball will probably land a little to the side, or a little later than you predict. That will be because you looked at the ball for only an instant, and your brain could not discern the speed or direction or location as accurately as you needed. Watching the ball a little longer before you close your eyes would improve your predictions. In the end, closing your eyes is never a good way to actually *catch* the ball, because at that point you want your estimate of where the ball is going to land to be accurate to a few inches, but if you just want a good indication of the ball's general movement, those first few moments of observing will suffice.

Object X is just like that ball being thrown. It is affected only by the force of gravity (the earth's gravity for the ball, the sun's gravity for Object X), so once we know where it is, how fast it is going, and the direction of motion, we know everything we need to know to be able to follow its orbit forever. Those first three hours that we had already seen, however, were like the very instant that someone threw a ball. If that's all you get to see, your estimate of where the ball is going will not be very accurate. We needed to keep our eye on the ball for a little more time before we knew the actual orbit of Object X.

In general, to understand the orbit of something so far away takes about a year's worth of precise observations. We couldn't wait a year. While we tried hard not to lose sleep at night thinking about someone else discovering Object X while we were still studying it, I would pick up the newspaper almost every morning with dread in my stomach. We were determined to wait long enough to write an accurate and thorough scientific paper on Object X, but we wanted to wait not a minute longer, for fear of being scooped. Wait until next year? No way.

Luckily, we didn't actually have to wait a year in the future. We could, instead, go back a year in the past. Many astronomers have taken many pictures of the sky over time, and perhaps we could find Object X there; by now, there were even online repositories of many of these images. Chad and I set to work in our separate offices across the hall from each other, probably looking at the exact same online pictures. I've heard stories of different parts of the same scientific team working in parallel on the same problem as a way of double-checking an important result, but I must admit, the fact that Chad and I were doing the same thing at the same time had nothing to do with double-checking. Looking back through the archive photos was simply so much fun that we both wanted to do it.

Here's how it worked, at least on my side of the hallway. First, I did the best calculation of where Object X was going and predicted where it should have been on a particular date a few months earlier. I then searched the archive for images at that position. Not surprisingly, there were none taken on the particular date I was looking for, but there were some taken a few weeks earlier. I went back and calculated the position of Object X for a few weeks earlier, and luckily, the position was right on that image. I downloaded the image from the archive and displayed it on my computer. The picture was full of indistinguishable stars. How could I know which was Object X? The only way to distinguish our discovery from the many, many stars in the sky was to see it move. But there was only one picture from that night, so there was no way to see it move. I could, however, go back to the archive and find a picture of that part of the sky taken a *year* earlier. Object X was moving, so a year earlier it would have been somewhere else entirely. I compared the pictures from the night when Object X was supposed to be there to the earlier pictures. It's easy on the computer; you just line the pictures up, press a few buttons, and the two pictures blink back and forth like a very short and repetitive movie. The two pictures were nearly identical. The stars and the galaxies had not changed at all over the year. But there, in the middle of the more recent picture, was a new starlike object that hadn't been there the year before.

That was what I was looking for; I still couldn't tell for sure that it was an object that was moving, but it certainly was one that hadn't been visible a year earlier. There are many things in the sky that can appear where they weren't seen before—stars that get brighter, stars that explode—so I didn't know for sure if this was our Object X or not. But if I assumed it was, I could calculate a little better where the object was going. With this more refined calculation, I could figure out where Object X should have been yet another full year earlier. I then restarted the whole process. Look for a picture in the right place; realize it is not quite the right time; revise the time; find the place; find an earlier comparison; look for something new. There it was! Right where I had predicted! I ran across the hall to tell Chad I had found Object X from a year earlier. He had found it a few minutes before me and was already looking for pictures from two years earlier. We were racing down the right trail.

We quickly followed Object X back for about three years, which was the limit of the data we could find online in the archive. While we were sitting in my office pondering what we might do next, Chad wondered aloud if perhaps Object X might be found on Charlie Kowal's plates. Ah yes. Charlie Kowal's plates.

Most of us have a blind spot, something we can't see even though it is right in front of us. Charlie Kowal's plates were directly in my blind spot. I knew about them but preferred not to think about them. Why? Kowal had, years earlier, proved that there were no planets out past Pluto. Since this information did not fit well into my view of the solar system, I chose not to think about it.

Charlie Kowal was an astronomer who had worked at Palomar Observatory in the 1970s and 1980s. He had decided to do something that no one had ever tried before: use the Palomar 48-inch Schmidt Telescope to find a planet beyond Pluto. At the time, Planet X was generally expected to exist (this was the 1970s, before the alleged evidence for the influence of Planet X on the outer planets had been thoroughly discredited), the 48-inch Schmidt was designed to cover large areas of the sky, and no one had mounted a serious search since Clyde Tombaugh. Thirty years later I would tell other astronomers about my search for planets, and they would frequently look at me critically and say, "Charlie Kowal did that thirty years ago, and he showed there was nothing there."

I had reasons for ignoring the critical astronomers. Kowal had, indeed, done almost exactly the same thing, but thirty years earlier he hadn't had computers around to do all of the searching for him. He had to look at each pair of photographic plates by eye and slowly search for anything that looked as though it moved from one night to the next. This was the job that I had calculated would have taken me forty years to accomplish, yet Kowal had done it all in something like a decade and in his spare time. I was banking on the fact that the only way Kowal could have looked at so much sky was if he went very quickly and paid attention to only the brightest objects on his photos. The fainter objects might actually be on his photos, but they would have slipped through his net. Many of my fellow astronomers weren't convinced by this argument and thought instead that I was off on a wishful-thinking fantasy chase. Chad's discovery of Object X made it clear that they were wrong in principle, and we now had a chance to see if they were wrong in practice. From published records, we found that Kowal had pointed the telescope directly at the predicted position of Object X on the nights of

May 17 and 18, 1983. If we could find Object X in those pictures, we would have a twenty-year-old position for Object X, and we would then know its full orbit exquisitely.

Kowal's photographic plates—and all of the other plates from fifty years of historic photographic work at Palomar Observatory—should have been stored in the airtight humidity-controlled halon-protected vault in the basement of the astronomy building next door to me on the Caltech campus. I went down to the vault, opened the lock, and peered inside, not sure exactly how I was going to find the specific photographic plates I needed among the thousands that were in there. The vault was in general disarray—no one had really used the photographic plates for a long while—but after letting my eyes adapt to the dim lights I could see that the place was laid out like library stacks, with the photographic plates in large manila envelopes arranged like books on the shelves, but by date rather than by author. I excitedly walked down the rows until I found 1983, and then I ducked into the aisle and looked up to where May should be, anxiously wondering what condition the plates would be in. But there were no plates. There was nothing. May of 1983—and several months before and after—were blank spots on the shelves, with little more than years-old dust. If the plates were misfiled, or perhaps had never been filed, the chances of my randomly coming across them in the vast vault were essentially zero.

That night I called Jean Mueller at Palomar. Jean had been involved with the 48-inch Schmidt Telescope for so long that I thought she might remember the Kowal plates and might know if they had ever been stored. She told me that, by chance, she was going to be down in Pasadena the very next day and would be happy to take a look. That day, the two of us went down to the vault, opened the door, and let our eyes adjust.

"I was down here a while ago, and I think I came across them," she said as she moved down the stacks. She quickly passed 1983.

"That's where they're supposed to be," I pointed out.

She ignored me, kept walking, and four or five rows later, turned left into an aisle between shelves crammed with manila envelopes full of photographic plates. She walked ten feet, turned right, reached up to the second-to-top shelf, pulled down an envelope, and said, "I think they might be around here someplace."

She wasn't quite right. She had put her finger on the plates from May 3, 1983—two weeks earlier than I needed. Our plates were about twenty-two inches to the right.

"How are you going to look at them?" Jean asked.

"Um, well, I was just going to look."

"You won't see a thing. Here, you'll want this." And she led me back to the front, where some decrepit equipment lay in disarray from decades of neglect. She handed me a light box—an ancient wooden tabletop enclosure with a slightly unsafe-looking power cord that, when plugged in, illuminated a photographic plate placed on top of it so that someone could examine it.

"We used to have a blink comparator"—the same sort of device Clyde Tombaugh had used to discover Pluto—Jean said. "Kowal would have used it on these plates himself. But I think that disappeared twenty years ago. You'll have to just look back and forth between the two plates and see what you see."

I put the envelopes containing the plates and the light box on an unstable rolling cart and brought them back to my office, almost knocking them down only once, when I had to push them over the lip and onto the carpeting in my building. I set the box on my table, gingerly plugged it in (carefully moving anything flammable from the vicinity), and flipped the lights on.

The plates were initially deceptive. They are rather heavy fourteen-inch-square pieces of glass kept inside large paper envelopes. When I pulled the first plate out of its envelope, I could see nothing at all except a few little marks apparently made by Kowal himself twenty years earlier, perhaps indicating candidate Planet Xs that he wanted to double-check.

Had the plates turned black with time? Was something wrong?

No, when I put the plate on the light box, I could suddenly see hundreds of stars, with large blank patches between them. I leaned over, my eye a foot away, and realized that each little patch of the sky that had looked blank itself contained hundreds of stars. And when I leaned all the way down and put my eye right up to the plate, I could see, it seemed, the whole universe in a single square inch, with countless tiny stars like glints from diamonds and myriads of swirling galaxies. And on this whole expanse of photographic plate, one of those countless tiny stars was, I believed, not a star, but was Object X and was moving from one night to the next.

I laid the plates from May 17 and 18 next to each other. On the two plates were countless stars, in precisely the same spots from one night to the next. Hiding amid them I was looking for one faint blip—Object X!—that jumped slightly between the nights. Only then, looking at the plates, did I truly realize the enormity of what Clyde Tombaugh had accomplished seventy-two years earlier by picking out Pluto from the stars. My job was easier. I knew roughly where to look on the photographic plate. I compared some of the bright stars to a modern star map, zeroed in on the approximate location, and boxed the area on both nights in felt-tip pen (very erasable from the glass surface). I then pulled out a hand-sized magnifier that was designed to ride over the top of the plates, and I started looking. I would look at one field of stars from the first night and try to memorize where everything was before looking at the second night. Was that star in a different place? Oops, no, I had just not noticed it before. How about that? Nope. Just a scratch on the plate. It took me thirty minutes to search one square inch of the photographic plates—about one-third of 1 percent of the total area—before I finally saw it. A tiny star was there one night but missing the next. And a second tiny star appeared the second night in a place where there was nothing the first night. I let out a scream, and then I forced anyone who walked down the hallway in my building for the next half hour to come in to look at the two spots on the photographic plates and see Object X as it had appeared in 1983.

It really was not surprising that Charlie Kowal had missed this one in 1983. It was a barely visible smudge that had taken me half an hour to find when I knew where to look and knew that there was something there to be found.

We now knew where Object X had been twenty years before, which meant that we could compute a very precise orbit for it. Just as important, we demonstrated that our hunt was not in vain. There might be more things out there that Kowal had not seen on his plates.

But first, we needed to get back to Object X itself. The orbit that we found was surprising. Object X goes around the sun every 288 years in an orbit closer to circular than even most of the planets, but it is tilted away from the planets by 8 degrees. Eight degrees might seem small, but compared to the planets it is enormous. What was Object X? How did it get its almost perfect but slightly askew orbit?

Today we still don't know the answer. We have elaborate theories of how the objects out in the Kuiper belt have been tossed around in their orbits by the giant planets, but all of this tossing both tilts *and* elongates the orbits. Tilted but circular? All but impossible. Finding out that something you have just discovered is considered all but impossible is one of the joys of science. It is an enormous clue to billions of years of the early evolution of the solar system. If only we knew what it meant. Eventually we'll piece together enough other parts of the story so that the peculiar orbit of Object X will suddenly make sense.

With the orbit and the position of Object X determined, we could finally try to answer the one question that had been burning in the backs of our minds. How big was it really? From the day of discovery we were convinced that it was bigger than Pluto. But we didn't actually know that for certain. Object X was so far away that, from our telescope, we couldn't tell that it was anything other than a point of light. It looked like a star; it was starlike, an asteroid by the literal meaning of the word, though that literal meaning had long ago been forgotten. Object X was bright, but all that "bright" means is that it reflects a lot of sunlight. An object can reflect a lot of sunlight if it has a shiny surface—because it is covered in snow, for example—or it can reflect a lot of sunlight if it has a darker surface but is really big. You would have the equivalent problem if you were on the ground and someone was signaling to you with a mirror high in the mountains. You wouldn't be able to tell the difference between someone with a small but highly polished mirror and someone with a larger but dirty mirror. Both would reflect the same amount of light in your direction. Both would appear as simple points of light from your distant vantage point.

There was, possibly, one telescope that could see the disk of Object X crisply enough that we might be able to directly measure its size. The Hubble Space Telescope orbits the earth high above the atmosphere and, now that the original defects in its mirror have been corrected, takes the sharpest pictures of anything around. Even the Hubble has fundamental limits—due not to defects but to the laws of physics—as to how tiny an object it can resolve, but I quickly calculated that if Object X was really the size of Pluto, then Hubble's newest camera, recently installed by visiting astronauts, would have no problem seeing the tiny disk and allowing us to measure its size.

To use the Hubble Space Telescope you have to submit a lengthy proposal—which is accepted only once a year—detailing what you would like to look at and why; then a committee of astronomers looks over all of the proposals and selects those they believe are the very best. The next due date for proposals was not for about nine months. The earliest we could possibly hope to get a picture from the Hubble was in about a year. We seemed to have only two choices. We could announce our discovery quickly, tell everyone that we thought it was likely bigger than Pluto, and then wait for a year to confirm. But our estimate of the size really was just an educated guess. What if our object was actually smaller than Pluto? We didn't want to have to be in the position to come back a year later and say that the thing we had called a new planet was actually *smaller* than Pluto after all. Our other option, though, was to wait a year so that we could announce the correct size when we announced our discovery. But we couldn't delay the announcement of our discovery for a year; someone else might find it in the meantime and not feel the need to know how big it was before making it public. And even if we *did* delay until after we got images from Hubble, we didn't think the secret would keep. Once the proposal was submitted, it would be read by dozens of people, and while proposals are ostensibly confidential, we were pretty sure that word would leak out quickly. Luckily, there was a third option.

It is understood that sometimes discoveries will be made that need pictures from the Hubble Space Telescope faster than the process will allow, so there is an official route by which you can appeal for data immediately. Even this route made me nervous. Many, many people would still be reading the request and learning about the object. So I went for an even more direct route. I sent a note to one person I knew who worked for the Hubble Space Telescope. I explained that we had just found something potentially bigger than Pluto and wanted to look at it with Hubble as soon as possible, but we were afraid to go through any of the official routes in case the information leaked. I attached a detailed proposal just like the one that I would have submitted, but requested that the fewest people possible know about it. I sent the note by e-mail and sat back to look at a few more images of the sky, but within about two minutes I had already gotten a reply: YES!

I quickly set to work trying to figure out the right time to target the Hubble. We wanted to make a very precise measurement of the size, so we knew we wanted to take the pictures just as Object X was moving close to a distant star to which we could compare it. I called up archival images of the sky, had the computer draw in the path that Object X was going to take through the stars, and looked for a good time. I found that in only three weeks the object was going to skim past a bright star; the timing would be perfect. I designed the precise sequence of pictures for the Hubble telescope to take and then sat back to wait the three weeks.

Normally that three-week wait would have driven me crazy, but I had a distracting trip planned. I was flying out to Hawaii to use one of the Keck telescopes—the largest telescopes in the world—to take a first really good look at Object X. Just as with any of the other great telescopes in the world, getting to use a Keck telescope requires writing a detailed proposal explaining what you will use the telescope for and why it is a good use of the time. As usual, the proposal is read by other astronomers, and then three to nine months later you might find yourself assigned to a particular night at the telescope. Unfortunately for us, again, we didn't know we were going to discover Object X ahead of time, so we couldn't have already written the proposal. Luckily for us, though, I had written a proposal to do something else entirely at the Keck—to study the moons of Uranus for evidence of icy volcanoes—so I was scheduled to be at the telescope soon after our discovery. One of the unspoken rules of being at a telescope is that once you are there, the night is yours to do with what you want. Yes, I had planned to look for icy volcanoes, but looking at Object X would clearly be a much more interesting and pressing use of the time.

The Keck telescopes sit atop the currently dormant summit of the giant Mauna Kea volcano on the Big Island of Hawaii. At nearly 14,000 feet above sea level, the summit looks more like the sterile surface of the moon than part of a fertile tropical island. The only sign of wildlife I have come across up there was a mouse who must have hitchhiked up in an equipment shipment and who lived on the crumbs dropped by astronomers or others working inside the dome. If the mouse ever got itself locked out of the telescope, it would find nothing to eat for miles around.

While the majestic Hale Telescope at Palomar Observatory looks like part spotless battleship, part elegant WPA dam, and part nineteenth-century high-rise, the monster Keck telescopes look like nothing but high-strung engineering projects. The dome at Palomar is mostly empty space, with the smooth outlines of the telescope truss looming high above in the darkness. The domes at Keck are the same size, but the mirrors on the telescopes are four times as big, meaning that the telescopes are so tightly crammed into the domes that there is nowhere to stand to even get a good perspective on what the telescopes look like.

If you take one of the elevators that goes midway up a dome and step outside onto the metal platform encircling the telescope, you can walk around and get some idea of the different components—white girders, sprawling wires and cables, massive industrial-sized cranes—and you will find yourself looking directly into one of the two biggest telescopic mirrors in the world. It's not one mirror, though; it is a bug eye of thirty-six smaller hexagonal mirrors all arranged into a much larger, almost circular hexagon looking back at you. The mirror itself, all combined, has a square footage only slightly smaller than the house that I lived in.

Later that night, when we pointed the telescope at the faint dot in the sky that was Object X, the mirrors would concentrate all of the light from that immense area onto a tiny spot about the size of the period at the end of this sentence. Our goal was to take that concentrated light and pass it through a system that acts as a prism, to spread the light out, and then look at the different components. By looking at this spread-out light—the spectrum—I hoped that I could determine what was on the surface of Object X.

I was scheduled to be at the telescope for two nights. I arrived in Hawaii a day early to begin to shift my body to a nighttime schedule and to do final preparations far from the distractions of home (including planning a wedding that was now only seven months away). I stayed up late at the observatory's headquarters refining calculations on the computer, and then I went to sleep with the hope that I would sleep until noon so I would be fresh for the long night ahead. Instead, I woke up before dawn. I tried to force myself back to sleep, but my mind was uncontrollably running through the plans for the night, how I would set up the telescope and instruments, what would be the best way to collect the most useful data possible. I gave up on sleep and walked over to the telescope control room to set up for the night.

The control room is arranged as a dense ring of desks around the center of the room, with an even denser ring of computer screens. At last count the room had something like twelve computer screens, all of which might be in use during the night. I checked the weather reports, the telescope reports, how things had gone the previous night. All of the nighttime staff from the observatory were still asleep, but there was plenty of preparatory work to do. At lunchtime, I walked to the shopping center to get some fresh Hawaiian poke from the grocery store.

Walked to the shopping center? No, there is not a shopping center on the desolate summit of Mauna Kea. I was in the little cowboy town of Waimea, only a couple of thousand feet above sea level and surrounded mostly by ranch land. To use the Keck telescope these days, astronomers rarely actually go up to the summit. Instead, we sit in the control room in Waimea and connect to the summit by a fast video and data link. We talk to the people there and control the instruments there, but we don't go there ourselves.

The first time I used a telescope like this while being in a control room miles away, I felt strangely disconnected from what was going on. I couldn't walk outside to feel the wind and humidity. I couldn't check for cloudy patches or impending fog. I couldn't hear the reassuring clanking of the dome and rumbling of the telescope. How could I do astronomy this way?

The answer is, nearly perfectly. Your brain doesn't work very well in the sudden oxygen deprivation of 14,000 feet. Combine that with lack of sleep, and efficient work is extremely hard. Fish-eye cameras pointing at the sky are better at seeing clouds coming and going than your eye will ever be. Wind and humidity gauges work just fine. And the video link is so seamless that you almost forget that you're not talking to someone sitting right next to you. Still, I always find it disconcerting when, on nights that I am working at the telescope and the sky at 14,000 feet is beautiful and clear and the humidity is low and we are collecting beautiful data, I think to look out the window and, outside the control room at 2,000 feet in Waimea, rivers of rain are being driven horizontally by gale-force winds.

Object X was going to rise above the horizon at about 8:00 p.m. I had finished setting everything up and was waiting anxiously to get started for the night. The crew arrived at the summit around 5:00 p.m., and we chatted over the video about the plans for the evening. When the sun went down, the big dome swung open and the thirty-six little hexagonal mirrors pointed together to begin collecting the light from my first target in the sky.

My first job was to do a very quick check of all of the systems. We swung to a nice bright star, focused the telescope, and put the light from the bright star down through the prism to see if everything worked. After a few minutes, the spectrum appeared on one of the computer screens in front of me. I typed a few commands to take a quick look; the spectrum of the star looked just as it was supposed to. I stored the data away to later compare it to Object X. Finally, it was time to find Object X. We turned the telescope in the right direction and took a picture to see what was there, and the picture that appeared a minute later on my screen showed that there were twenty stars more or less where I expected Object X to be. Which one was it? I knew how to find out: It would be the one that moved. We did a little more calibration, and then twenty minutes later we took another picture. At first glance, the picture looked precisely the same, but I lined up the two pictures on the computer screen and blinked back and forth between them. Nineteen of the twenty stars reappeared in exactly the same place. One of the stars had shifted slightly. It wasn't a star. It was Object X.

Though we had been studying it and tracking it for more than a month now, my first view of Object X through the giant Keck telescope—or at least on the computer screen twelve thousand feet below the giant Keck telescope—still amazed me. I was about to get the first peek at the composition of something that might be bigger than Pluto, something that only a handful of people on the planet even knew existed. I shifted the telescope slightly to direct the light of Object X into the prism, and we were ready. Though Object X was the brightest thing beyond Pluto that had ever been seen, it was still faint. Even with the biggest telescope in the world, we had to collect a large amount of light before we had enough to be able to make a sensible analysis. We stared at Object X all night long, stopping every once in a while to be sure that the light was indeed going into the prism. I watched the data come in and obsessively checked the weather reports. Everything went perfectly. No clouds, no fog, no telescope malfunctions. Everything went so perfectly that it was, to be honest, an incredibly tedious night. I occupied myself with loud music, junk food, double-triple-quadruple-checking that everything was going perfectly, and speculating about what I might find.

The sky began to brighten with the rising sun at around 5:30 a.m., and I finally made my way back to my little room. I slept until almost 11:00 a.m., went back to the control room, and again began preparing for the night. The second night was almost exactly like the first. I went to sleep around 6:00 a.m., got up the next day at 10:30 a.m., and was on a flight back to LAX by 1:00 p.m., confident that I had collected exactly the data I needed.

Two nights at the Keck telescope will provide weeks' or even months' worth of data to pore over. Though totally exhausted, I got started on the five-hour airplane ride back home, trying to use all of the pictures and data to create one coherent view of what we had seen. First, I had to carefully remove any effects that were caused by the telescope or the prism or the earth's atmosphere rather than by Object X itself; second, I had to figure out what we were seeing; and third, I had to figure out what it all meant.

It quickly became clear that we were seeing dirty ice. Perhaps that should not have been a big surprise for something so far from the sun. Ice was supposed to be one of the main components of Pluto, too, and it was on the surface of almost all of the big satellites of Jupiter, Saturn, Uranus, and Neptune. But in addition to the dirty ice, there appeared to be something that looked like frozen methane. Methane would perhaps not be surprising to find on the object's surface, since it is one of the main components of the surface of Pluto, but it had never been seen anywhere else in the Kuiper belt, and the signature of methane was not overwhelmingly convincing. If methane was there at all, it was in extremely small amounts. A few years later another astronomer would suggest that perhaps there was no methane at all on Object X, but that what I thought looked like methane was actually evidence for the same icy volcanoes on Object X that I was supposed to have been looking for on the satellite of Uranus to begin with.

The methane on Object X (and it was methane, after all) never made sense until years later, when Emily Schaller, a graduate student of mine working on a Ph.D. dissertation about the methane clouds on Titan, walked into my office with an idea for why Titan and Pluto both had methane. Her final explanation was deceptively simple and explained not just these objects but the rest of the Kuiper belt as well. Object X, it turned out, formed with methane—as did Pluto and Titan—but Object X was just a little too small, so that its gravitational pull was not quite strong enough to hold on to the methane forever. With the Keck telescope we were seeing the very last remnants of frost on a cold, dying world.

While I was still working to understand the data from the Keck observatory, the Hubble Space Telescope snapped its sequence of pictures and transmitted them to the ground, where they were sent to my computer in Pasadena. Because the Hubble is totally automated and you design the entire sequence ahead of time, you can very easily lose track of when the telescope is actually looking at your target. The Hubble pointed at Object X on a Saturday, as I was having a housewarming party to welcome Diane as a new resident of my—now our—home. The house, with a square footage only slightly larger than that of the Keck telescope, was a bit of a tighter fit now. I didn't make it to work until Sunday afternoon, after a long cleanup from the party. The new data would immediately tell us how big Object X was. Much bigger than Pluto? Only a little bigger? A tad smaller? When I first opened up the file that contained the image, I immediately closed it and double-checked what I was looking at. Clearly this was not Object X, the object potentially larger than Pluto—how could it be? But yes, the tiny dot that surely couldn't be the tenth planet was, indeed, Object X. Object X, in the end, turned out to be only about half the size of Pluto.

How could this be? How could we have turned out to have been totally wrong? The answer, in a single word, is albedo. Albedo is a measure of how reflective something is. Freshly fallen snow has a high albedo, while coal or dirt has an albedo that is quite low. No one really knew what albedo to expect for things in the Kuiper belt, but back when the first object was found, everyone assumed that they were dark—as dark as coal or soot or ash. When we see an object out in the Kuiper belt, all we see is sunlight reflected from the surface. If that surface is dark and doesn't reflect much light, the object needs to be big to reflect a lot of light, but if the surface is icy or shiny for some reason, it can reflect just as much sunlight while being smaller. It turned out that Object X was not as dark as coal or soot or ash; it was more like ice with a bit of coal or soot or ash thrown in. It was shinier than we'd initially guessed, meaning that it was smaller than we'd thought.

I was disappointed at the time, but only a little. We were just getting started, and we had planets in our sights.

Now that we finally knew how big it was—no planet for sure—it was time to give Object X a more dignified name. There are rules, decided upon by the International Astronomical Union, for the naming of most everything in the sky. Craters on Mercury have to be named for deceased poets; moon of Uranus are named for Shakespearean characters. For this type of object in the Kuiper belt, the rules said that the name had to be a cre-ation deity in a mythology. After some quick thought, Chad and I decided that we should move from Old World mythologies, which have been traditionally used, to New World mythologies, in honor of where Object X was found. We even thought we might try to preserve the *X*. If you're looking for New World mythologies and names that begin with *X*, you can do no better than the Aztecs. They were fond of *X* names—Xiuhtecuhtli is one of my favorites—but none of those felt quite right, or quite pronounceable. A little more Internet searching brought us to consider more local deities. Object X had been found at Mount Palomar, which is surrounded by Native American tribal reserva-tions. Did the Pala tribe have deities? The Pechanga tribe? What gods did they worship in earlier days? We searched the Internet but couldn't find any; our search brought up only early-eighties entertainers who were currently playing at their massive Harrah's casinos, whose Las Vegas–style lighting is slowly ruining the view of the sky above the telescopes on top of Palomar. But we did find something even more local: The Tongva tribe, mostly known as the Gabrielino Indians because of their proximity to and assimilation into the San Gabriel Mission, had long been the inhabitants of the Los Angeles basin. In their mythology, the world was begun when their creation force—called Kwawar— sang and danced the universe into existence. It occurred to us, though, that there were actual members of the Tongva tribe around and that we really should ask their permission first.

We didn't know anyone in the Tongva tribe, but Chad went to www.tongva.com, found a phone number, and called it. The chief answered. Chad said something like, "Hi, I'm an astronomer from Caltech, and we just discovered something big in this region of space called the Kuiper belt and were hoping to name it after a Tongva creation myth and wanted to talk to you about it," at which point the chief probably thought there was a pretty good chance that Chad was a lunatic rather than an astronomer from Caltech. Perhaps to hedge his bets, or perhaps just to get rid of Chad as quickly as possible, he gave the name of the tribal historian and chief dancer, who would be a better person to talk to about such matters.

Chad made the next phone call. After Chad convinced the tribal historian that he was not a crazy person but was indeed an astronomer who had found something half the size of Pluto that needed a name, the Tongva agreed that Kwawar—or rather Quaoar, their preferred spelling—was the appropriate name.

The correct pronunciation of Quaoar sounds like Kwa-o-ar, with a very soft *W* sound and a bit of a Spanish roll to the *R,* no doubt a product of the mission days. Simply saying Kwawar works fine, too. But when we picked the name, it didn't occur to us that if you didn't see it spelled Kwawar originally, as Chad and I had, the English language doesn't give many clues on how to pronounce the word correctly. No word in the entire English language has that particular combination of four vowels: uaoa. People trying to pronounce it tend to start with the Q and then quickly trail off into nothingness.

With a name in place, we were now ready to announce to the other scientists and to the world what we had found. A large international meeting of astronomers was taking place in Birmingham, Alabama, just two hours from my hometown, and we decided to make the announcement there. Chad submitted a paper with the innocuous-sounding title "Large Kuiper Belt Objects." In his talk, he discussed everything that we had learned: Quaoar's oddly circular yet inclined orbit, its diameter about half the size of Pluto's, its icy surface. All of the questions, though, had nothing to do with Quaoar. Most of the inquiries from the press that day and over the following weeks never even mentioned Quaoar itself. They just wanted to know one thing: what did this discovery mean for whether or not Pluto was a planet?

What, indeed? Even as more and more objects in the Kuiper belt were being found, Pluto still stood out as being significantly larger than any of the rest—but it was larger than Quaoar by only a factor of two. Was that enough to doom Pluto? In many ways, the answer was clearly yes. If after only nine months of looking, we could find something half the size of Pluto, how much longer would it take to find something the size of Pluto? We figured it was only a matter of months. For the confirmed Pluto fans, finding something smaller than Pluto meant nothing; Pluto was still the biggest, and thus they could go on calling it a planet. Yet it seemed that perhaps Pluto, while not yet dead, was on its deathbed. As *The Birmingham News* quoted me as saying later that day, Quaoar was a big icy nail in the coffin of Pluto as a planet.

The week after we returned from Birmingham, Caltech threw a black-tie dinner to announce the kickoff of an ambitious fund-raising campaign. Many of the people at the dinner were donors who had been with Diane on one of her many Caltech travel-study trips around the world. Having just been in the newspapers a week earlier for the discovery of Quaoar, I was a minor celebrity at the party. Being engaged to Diane, though, made me a major celebrity.

I spent the evening in a conversational loop: "You're the person who discovered that thing out past Pluto?"

Yes, indeed.

"I want to introduce you to my friend—hey, do you know Mike Brown? He's the guy who discovered the thing past Pluto."

"Sure, I know Mike; he's the guy who is engaged to Diane Binney. Hey Mike, I want to introduce you to my friend—hey, do you know Mike Brown? He's the guy who is engaged to Diane Binney."

"Sure, I know Mike Brown—he's the guy who discovered that thing out past Pluto. Let me introduce you to a friend who is really interested in planets. . . ."

Chapter Six

THE END OF THE SOLAR SYSTEM

Even today I spend much of my time exploring the outer edges of the solar system, looking for little worlds that have never before been seen, wondering what else is out there on the outskirts of our solar system. Someday I will have looked everywhere that the telescopes I have are capable of seeing, and then I guess I will have to declare that my days of exploring are finished.

It will be nice to finally stop fretting every night when I see a few clouds in the sky as the sun goes down, or when the moon is nearing full and I know that the section of sky we wanted to cover this month is not quite done. It might be nice to wake up in the morning and see red-tinged cumulus clouds beautifully strewn across the L.A. basin and not have to wonder what we missed last night. And even though the computer does most of the hard work of looking at all of the data and finding the things that move, something always goes a bit wrong and I am always fixing a little bit of computer code or making slight improvements. The computer even sends me text messages on my cell phone when something goes really wrong. More often than not, it seems, trouble occurs on Saturday mornings while I am sitting drinking my coffee.

Still, the fact that on any morning I might walk into my office and see something moving across the sky that no one has ever seen before, something bigger than anything found in perhaps a hundred years, adds an element of excitement to my life. I will be sad to be done, and what will I do after that?

I did almost quit once, a little more than a year after the announcement of Quaoar. I thought, at the time, that we had reached the end of the solar system.

Chad had moved back to Hawaii by then, eventually to marry, buy a house on the rainy, steamy, jungly northeast side of the Big Island, and work on telescopes. He and I (though, really, mostly he) had spent two long years staring at the sky night after night, and by the end of the two years we had covered 12 percent of the whole sky. While this might not seem like a huge amount, this time we really had covered a wide swath of the parts of the sky where we expected anything big to be. If we looked farther north or farther south, we would be looking away from the region where all of the planets are. The only things that we would find in the regions farther north and south would be things that went around the sun in orbits even more tilted than Pluto's. The chances that something like that was out there seemed remote.

I don't mind taking bets on remote chances. Perhaps you could have said that our chances of finding something as big as Quaoar were remote, too, but there it was. The chances I would meet the person that I was going to marry in the basement of the 200-inch Hale Telescope were even more remote, but by now Diane and I had been married almost six months. Remote chances lead to good things, as far as I can tell.

So in the fall of 2003, just as Chad was leaving and our two-year project to use the little telescope at Palomar to scan the skies for planets was ending, I began a new project about which I was quite excited. I was going to use the same telescope to scan the skies for planets. For the third time. This time, though, I wasn't going to concentrate on the most probable places, I was going to concentrate on some of the least probable. The project was going to be even better than before, too, because other astronomers had become interested in using the telescope to look at vast areas of the sky for very rare quasars flickering at the edge of the universe, and they had built an even bigger camera—the biggest astronomical camera in the entire world!—to look at even bigger areas of the sky at once. This seemed, at least at first, like great news for us. We would sweep through the unsearched regions of sky faster than ever before.

Right before Chad moved back to Hawaii, he modified all of the computer programs he had written over the previous three years so that they would work with this new supercamera. He automated everything as much as possible so that the project could continue in his absence. I was a little nervous about this, because it meant that I was stepping in to be the one in charge of the night-to-night workings of the project. I had been letting Chad take all of the major responsibility for years now, and in that time, I'd had many other projects going on to worry about and spend my time on. But things looked good. It looked as though with just a little bit of babysitting from me everything would run smoothly, the skies would be ours, and I could keep my day job.

The new camera arrived about a month after Chad left, and it spent its first night taking pictures of the sky. At the end of the night, I set Chad's computer programs to search once again for distant planets, for things that were moving in the sky. The computer worked all day long, as I carried on with all of the nonplanet-searching projects that were supposed to be occupying my time. Finally an automated e-mail informed me that the program was done. I opened up the file to see if the program had found anything. It had! Not only had it found things moving in the sky, it had found thirty-seven thousand of them!

My heart sank.

There could not possibly be thirty-seven thousand real moving objects in pictures from that night. In fact, I now know that there was precisely one.

The computer was confused. But it was not Chad's program that was the problem, it was the fancy new camera. To make the biggest astronomical camera in the entire world at a price that was not astronomical, the builders had had to compromise a bit on quality. One of those compromises had led to an incredible number of smeared spots, dark blemishes, light dots, black streaks, and bright blots showing up in each and every picture of the sky. The computer doesn't do a good job of distinguishing between bright blots or light dots caused by the camera and those caused by something actually in the sky. Those thirty-seven thousand moving objects were almost all camera junk.

I had not expected the computer or the camera to be perfect. I had anticipated that every morning I would have to look through *some* of the pictures to sort out the real objects from the fake ones. I had even taken the time to write a quick computer program to make this sorting extremely efficient; I could simply sit at my computer, press a single button, and a little postage-stamp-sized bit of the pictures from the previous night would appear on my screen. Three images would blink through in succession, and by eye, I would quickly be able to see what the computer had thought was moving. The eye is really good at finding the computer's mistakes or verifying true finds. After some practice, I could look at perhaps as many as twenty different candidate objects in a minute. But to look at thirty-seven thousand would take me thirty hours straight for every one night's worth of data. This was potentially a disaster.

I sent an e-mail to David Rabinowitz, an astronomer at Yale University, describing the problems. David had helped build the new camera and had joined Chad and me as the third member of our planet search team; if anyone knew any clever solutions to the problem, it would be David. He quickly responded: There was nothing that could be done to fix the camera's problem.

The only plausible solution I could think of was to somehow make the computer program much, much smarter. But Chad was on to a new job and new responsibilities and couldn't spend the next two years writing new computer programs the way he had for the previous camera. And even if he *was* working on this project, I couldn't think of an obvious way to make the computer program smarter. Everything that I could think of doing to get rid of the thirty-seven thousand camera-junk objects had a chance of getting rid of the real objects, too.

There was one solution: I could quit. Shut the project down. Declare an end to the solar system. In fact, it almost seemed like a good idea. Our chances of finding new objects were remote. The effort to find them was going to be extreme, if not impossible. If ever there was a time to cut our losses, now was it.

I needed a second opinion. I walked up the road to my favorite café with Antonin Bouchez, one of my graduate students at the time and someone whose opinion I greatly trusted.

"I'm done," I told him. "We've looked at enough sky; if there was anything else out there, we would have seen it by now. The new camera is low quality, and I don't think there is really any way to move forward."

I laid out all of my reasoning. I outlined the regions of the sky we had covered. I talked to him about the very slim probability of finding anything else. I showed him data on the new camera.

"You're crazy," he said.

"No no no," I told him. I went through the arguments again. Look at the problems with the camera! Look at how well we've already done with the sky!

"No, really, you're crazy."

We drank more coffee. I described how I believed the solar system was laid out and why it now seemed clear that there was nothing larger than Pluto out there to be seen. And thirty-seven thousand moving things to look at in one night? Impossible!

"Do you really believe there's nothing else out there?" he asked.

"I do," I said.

"So how are you going to feel when you pick up a newspaper one morning and read about someone discovering something right where you didn't look?"

I was reaching for the coffee again but stopped short. "Uhhhhhh. But it's not going to happen since we've reached the end of the solar system."

"What if you're wrong?"

What, indeed? Ten years earlier almost no one had thought that there was anything to be found beyond Pluto at all and that anyone spending all of his time looking was crazy. Even just two years earlier almost no one had thought that something as big as Quaoar would be found and that anyone spending all of his time looking was crazy. I hadn't bothered believing what most people thought back then, so why was I bothering to believe what most people thought now?

"Do you really know there is nothing else out there?" Antonin asked again.

Well. Okay. No. I really didn't.

"Then why exactly do you want to quit?"

Because it was going to be hard work. Because I didn't have help anymore. Because I wasn't certain I'd be able to pull it off alone. Because I had been working on it for a couple of weeks and had hit what felt like insurmountable roadblocks.

Looking back from a perspective of more than half a decade later, I think of this conversation as being as momentous as the moment when Diane walked through the door of the 200-inch Hale Telescope that first time that I saw her and my life irrevocably changed. A decade of floundering had ended that moment. This time the floundering had been for only a few months, but I had been floundering nonetheless. I can now even identify what the problem was, though I couldn't have done so at the time. My biggest problem was not that the camera had specks or that the software was not up to the task. My biggest problem was that I had let myself become a normal person instead of an astronomer. I was believing what most people thought, because "most people" now included me.

When I hired Chad and set him to work, he was so good at it that I had spent most of the previous year or so enjoying my life. Most nights I even left work at an almost reasonable hour and went home and made dinner for Diane—and the nights I didn't were usually because *she* was working late, not me. In the year before the new camera had been connected, Diane and I had married, gone on a monthlong honeymoon to South America, been on vacations, fixed up our little house. In short, we were behaving like normal people. I had never quite behaved like a normal person before.

I could do all of this because Chad was hard at work every night scanning the skies. And he periodically let me know how it was going. But, really, I didn't know much about the details of what he was doing.

Now Chad had moved on to a new job, and I was left with a big complicated system that was suddenly mine alone. And a major component of the system had just changed, and everything needed to be fixed, and no one knew how to do it.

Antonin and I were still drinking our coffee. "Keep looking," Antonin said. "How could there be nothing left to find?"

I had used that same argument myself. How could there be nothing left to find? How could this really be the end of the solar system?

I drank more coffee. I stared into space. How would I do it? There was no way I could get someone up to speed quickly enough to keep going. We were still scanning the skies every single night. I didn't have the time to wait months or years for someone new to come on board to get things going. I needed someone right now.

And then I thought of someone who was actually pretty good at this sort of thing and even knew a bit about it already. Me. It would mean an end to being normal, to going home most nights and cooking dinner, but it would mean that the solar system didn't have to end.

I finished my coffee, and Antonin and I headed back toward campus, but I took a quick detour to Diane's office. She was between meetings. I told her about the problems and the 37,000 objects and about the solar system that I didn't want to end and how the only solution was to start doing all the work myself. She looked at me, smiled, and said, "Go find a planet."

In the end, the solution to what to do with 37,000 objects in one night turned out to be deceptively simple: I let it go. After a few more nights of collecting data and finding 33,000, 50,000, 20,000, and 42,000 objects, patterns began to emerge. Almost all of the camera junk turned out to be in a few places on the images. If I just threw some parts of the pictures away, ignoring what was there, then everything else was suddenly manageable. That meant, of course, that if something real was there I had to throw it away, too. But it was a price I was willing to pay. I finally settled on throwing away about 10 percent of the sky to get rid of 99.7 percent of the camera junk. For the first night's worth of data, I went from 37,000 potential objects to look at to about one hundred. I could handle one hundred.

I spent a couple of long days and nights at my computer going through the two months' worth of pictures that had accumulated while I had been figuring out what to do. Of that very first night's one hundred objects, one turned out to be a real object out there in the Kuiper belt. It wasn't the biggest we had ever seen—it was only about one-third the size of Pluto—nor did it really distinguish itself in any other major way, but there it was, a tiny little needle that I had found by throwing away only 10 percent of the haystack.

On one of those late nights when I was sorting through recent data, I found a bright Kuiper belt object; and then five minutes later, one more; and then five minutes later, a third. Again, they were not the biggest or the brightest objects, but it was clear we were in business. I let out a little shriek, which caused Emily Schaller—my graduate student who was working on Titan's methane clouds—to stick her head in my office to see if everything was okay.

The objects I found didn't look that special—the postage-stamp-sized picture just showed a single faint point of light moving slowly across a patch of sky full of stars. I don't know if it was the fact that no one had ever seen this little world before, that something in the sky was moving, or that this thing I was seeing was near the edge of the solar system, but each discovery of one of those moving dots on my screen gave me a charge of adrenaline and a jolt of excitement. Even today, when I see one I want to grab whoever is in the hallway and sit him or her down in my chair and point. Look!

Over the next months, I barely kept my head above water. I was refining the software, making sure the telescope looked in the right places, flipping through a hundred or more images every morning, and still spending most of my time on the class I was teaching. My class that fall was called The Formation and Evolution of Planetary Systems, which taught graduate students current thinking on how the solar system is constructed. A lot of the time, the lectures focused as much on what we don't know as on what we do. One of my favorite lectures was titled "The End of the Solar System"; it was where I got to talk about my own work in relation to the rest of the solar system. One of the mysteries I had been working hard on for the past few years was why the solar system seemed to end so abruptly. Yes, it continued on farther past Pluto than anyone had initially guessed, but about 50 percent farther than Pluto's current distance from the sun everything came to an exceedingly abrupt end. Nothing had ever been found beyond this distance, and no one knew why. It is a mystery that still dogs and excites me today. I've gotten pretty good at ruling out almost any idea that anyone ever has. But I am just as good at ruling out my own ideas.

I'd prepared my lecture more quickly than usual the morning of November 15, 2004, since I knew the subject intimately. I had a few extra minutes before class, so I decided to look at the images from the night before. As usual, almost everything that showed up on my screen was an obvious mistake the computer had made. But after a few minutes, I stopped my quick flipping through images, because I had found one that confused me. A faint object moving slowly across my screen—more slowly, in fact, than anything I had ever seen before.

The speed with which an object moves in our pictures is directly related to how far away it is, in precisely the same way that when you're looking sideways out the window of a speeding car, the things nearby zoom by quickly while the mountains in the distance appear to be just barely crawling along. The fact that this thing that I was looking at was moving at about half the speed of anything else I had ever seen meant that, if it was real, it would have to be twice as far away as anything anyone had ever found.

Most of the time when I find a real object, I know it right away. Most of the time, the thing that I see moving across the screen is unmistakably real. But this one was moving so slowly and was so faint that I couldn't decide whether or not it was real. It could have been just a series of slight smudges that had coincidentally lined up but meant nothing. If you look at the sky for long enough, you're bound to find such things. But what if it *was* real? What would it mean to find something so far away? I didn't have any more time to think, because it was time for my class.

I gave my normal lecture. But at the end I couldn't resist. After I told my students all about what we understood to be the edge of the solar system, I stopped, looked up, and added, "Maybe." I told them that I had perhaps just found something that had changed all of that. But I wasn't sure. And I would keep them posted.

I went back to my office and sent an e-mail to Chad and David. I tried to downplay the potential discovery:

Subject: amusement
  I just found something that, if real, is at 100 AU. Wouldn't that be fun?

Something at 100 AU—a hundred times the distance from the earth to the sun—would be more than three times the distance of Pluto and well beyond anything ever found in the Kuiper belt. Chad wrote back almost immediately:

If that one is real I'll be buying the champagne.

Chad and I eventually drank that champagne. We were sitting on a beach on the Big Island of Hawaii, with the sun setting over the ocean in front of us, a pig roasting in a pit behind us. By entirely appropriate chance, Antonin, who had convinced me not to quit my search for a new planet, was there, too. We raised our plastic cups to an unending solar system.

. . .

This was it; something so far away that we could nonetheless see had to be big—almost certainly bigger than Pluto. It's true we had been fooled by Quaoar at first—since it had had a much shinier surface than we had anticipated and was thus unusually bright without being as large as Pluto—but even if this new object had a surface as shiny as Quaoar, it would still have to be bigger than Pluto. Because it had been so elusive, we gave this new object the code name Flying Dutchman. The Flying Dutchman is, of course, the ghost ship of folklore that can never go home and is instead destined to sail the seas forever. We had no idea at the time what an appropriate name this was.

Since the Flying Dutchman—or Dutch, for short—was farther away than anything anyone had ever seen before, it certainly seemed to be part of a new, previously undiscovered part of the solar system. But I knew that there was another possibility. Even though Dutch was currently far beyond the Kuiper belt, it could still really be part of it. Sometimes objects in the Kuiper belt come a little too close to Neptune and get flung out onto long, looping orbits.

We do the same sort of flinging whenever we want a spacecraft to get somewhere in a hurry; we send it by Jupiter first to get a slingshot off the planet. The trick is to aim the spacecraft *almost* at Jupiter. The spacecraft gets closer and closer to Jupiter and is pulled faster and faster by the gravity of the giant planet, and then it just misses, skims the clouds, and now zips along at high speed toward its final destination. Jupiter is so massive that it has enough gravity to give an object a slingshot that will take it clear out of the solar system. The Pioneer and Voyager spacecraft went past Jupiter, took pictures, got the slingshot, and will never be seen again. Neptune, however, is too small to give a strong enough slingshot to propel something out of the solar system, so when it tries, the objects always come back. Many objects in the Kuiper belt thus have orbits that take them close to the orbit of Neptune but then much, much farther away from the sun. These objects have been called "scattered" Kuiper belt objects, as Neptune appears to have scattered them to those looping orbits.

Only small things get scattered. The large planets are on nice circular orbits because there is nothing big enough to kick them around. The objects in the Kuiper belt—including Pluto—have tilted, elongated orbits because they are too small to resist the bullying of Neptune. Dutch could well have been a scattered Kuiper belt object rather than something on a circular orbit like a planet. Maybe we just happened to be seeing it so far away because it was at the most distant point in its scattered orbit and would soon be making its way back toward the sun to show that it really belonged to the Kuiper belt region. Its orbit would be a clue to its potential planethood.

As we had with Quaoar before, we eagerly looked for pictures of Dutch that had been inadvertently taken by previous astronomers. Dutch was much fainter than Quaoar had been, so there weren't nearly as many on which it showed up, but after a few days of careful searching we found it back a few years, which was enough to calculate what sort of orbit it had.

What was it going to look like? Circular, the way the orbits of massive planets should be? Scattered, like the orbits of many of the other smaller objects in the Kuiper belt? At first it was hard to tell. Although it is true that you need to figure out only where an object is and how fast it is moving to know an orbit, Dutch was so far away and moving so slowly that every time we measured it we came up with a slightly different answer. First we thought its orbit was circular; then we thought it was moving in a straight line and not even in orbit around the sun (that would be a first!). But after more care and measurement, we finally got the answer: Dutch was definitely not moving in a circular orbit, and it was definitely not moving in a straight line. The orbit was extremely elongated. So was Dutch at its farthest point in its orbit and moving inward like a normal scattered object would? No: just the opposite. It turned out that Dutch was at almost its closest point and moving *outward*. And its orbit around the sun appeared so elongated that it was going to take eleven thousand years to go all the way out and come back in again. It was the most distant object that humans had ever seen in the solar system, but it was eventually going to be even ten times farther away. Nothing was supposed to act like this in the solar system. It was neither a normal-seeming planet nor a normal scattered Kuiper belt object. There was nothing like it known anywhere else in the universe.

It's sometimes hard to picture all of these orbits and what they mean. So try this. Take a sheet of copy paper, a pencil, and a quarter (or just follow along on the diagram on the next page).

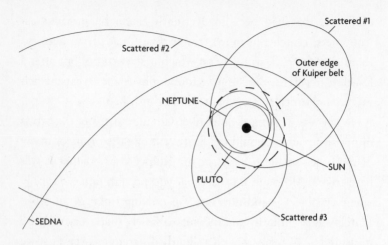

Put the quarter in the middle of the paper, trace its outline, and put a little dot at the center of the circle you have just drawn. This little dot is the position of the sun, while the outline of the quarter is the nice circular orbit of Neptune. Inside this circle is everything in the solar system that was known until the moment that Pluto was discovered in 1930. If you would like to put Pluto on your drawing, put your pencil at the four o'clock position of the circle of Neptune's orbit and now draw an oval that starts and ends there, but while it goes all the way around the sun it reaches a distance almost but not quite twice the diameter of Neptune's circle from the sun at the ten o'clock position (okay, if you're being precise, get out your ruler and make Pluto go 1⁹⁄₁₆ inches from the center of your circle). Now you can draw the outer edge of the Kuiper belt: Sketch a rough circle all the way around the sun at the farthest distance of Pluto. Finally, shade in all of the space between Neptune and this outer circle. Now it is time to add a few scattered objects. Place your pencil at, say, a point halfway inside your Kuiper belt at the eight o'clock position. Now draw an oval all the way around the sun that starts and ends here but gets to a distance two or three times farther by the two o'clock position. Feel free to draw as many scattered objects as you like, just always make sure to start and end in the middle of the Kuiper belt before zipping off to the edges of the solar system.

Now you will need to draw Dutch. Draw a little dot about three times as far from the sun as the orbit of Neptune at, say, the one o'clock position (again, you precision freaks, put that dot precisely 2⅜ of an inch from the sun). You're forgiven if, at this point, you would like to now draw an oval around the sun by coming into the Kuiper belt before going back out to your one o'clock position. But don't do it. Dutch never gets much closer to the sun than where you drew it. Instead, take your pencil and draw an oval around the sun that starts and ends at the position of Dutch; but at its most distant point, at the seven o'clock position, the oval needs to be farther away. How much farther? Almost 33 inches—three times the full length of your 8½-by-11-inch paper! Dutch never touches the Kuiper belt. It never comes close to Neptune. And it spends most of the time so far away from the comparatively tiny region that is the Kuiper belt that from Dutch, the sun would be just an extrabright star in the sky. There is nothing else like Dutch.

Now take your paper and put it in a safe place for later study. It will be on the final exam.

Even though nothing like Dutch had ever been seen before, I had an idea about what it was immediately.

One of the benefits and joys of teaching a comprehensive class on something like The Formation and Evolution of Planetary Systems is that you learn an awful lot about the formation and evolution of planetary systems. Much of my day (and late nights and early mornings) is spent with the concepts that I want to teach spinning in my mind. I see and continuously rearrange the outline for whatever is my next lecture as I am lying in bed or driving home or cooking dinner or eating breakfast. I mentally go through all of the connections and logic and calculations to make sure they make sense.

On the very day I realized that Dutch was unlike anything else known in the universe, I was mulling over my next class lecture, which was about the origin of comets. Dutch had an orbit *almost* like that of a comet. Comets are tiny balls of dirty ice that come from far out in the solar system, quickly swing by the sun, and return again. Dutch did the same, but it never came nearly as close to the sun as a comet, nor did it go nearly as far away from the sun as a comet. Comets acquire their distinct orbits through a complicated dance with giant planets and passing stars, and—I quickly calculated—Dutch never comes close enough to any of the planets to be a partner in any such dance. But while working on my lecture for the day, I quickly realized that Dutch could have acquired its odd orbit if, when the sun was born 4.5 billion years ago, the sun was not an only child but, rather, simply one in a litter of many stars. Before all the other stars went their own ways, they could have pushed Dutch around and put it exactly where it is now. Astronomers had speculated about such things for decades and had argued back and forth about whether it was true, and I had just found the thing that was going to answer all of those questions for good.

Discovery is exciting, no matter how big or small or close or distant. But in the end, even better is discovering something that is capable of transforming our entire view of the sun and the solar system. Dutch was not just a chunk of ice and rock at the edge of the solar system. It was a fossil left over from the birth of the sun. And as surely as a paleontologist can take a fossilized bone of a *T. rex* and learn what the earth was like 70 million years ago, I was pretty sure that we could examine this fossil in space—this object that could have been put in place only near the very moment of the sun's birth—and learn more about the sun's earliest childhood than ever before.

That class was the most astounding I have ever taught. I carefully explained the steps and the calculations that show why comets are where they are and why something like Dutch—which they still didn't know about—could not possibly exist, at least given the standard picture of the formation of the solar system. And then I showed them Dutch. Finally, I went through the same calculations but now with different conditions 4.5 billion years ago and showed that it would lead precisely to things like Dutch. QED. The students in the classroom dutifully took their notes, probably thinking nothing more profound than whether or not this would be on the final exam. Of course it was.

After Quaoar, I learned an important lesson. Names should be pronounceable. In the end, when it was time for a real name for Dutch, I settled on Sedna. Sedna is simple and easily pronounced, and has a serene sound to it.

The name Sedna comes from Inuit mythology. Since Dutch was so far away from the sun and was the coldest object anyone had ever seen in the solar system, I was looking for a name from an appropriately cold region. I quickly settled on Inuit as the closest polar mythology to my home in Pasadena. Sedna is the goddess of the sea. She lives in an ice cave at the bottom of the ocean, which seemed pretty cold to me. Plus the name has only two vowels—and they are not in a row. She does not, however, have a pleasant backstory.

In Inuit mythology, Sedna was a young girl who refused to marry any of her many suitors. Her father finally forced her to marry a mysterious stranger who couldn't be seen beneath his cloak. The stranger was a raven, who took the girl back to his nest. Her father finally heard his daughter's screams and, filled with remorse, crossed the sea in his kayak to rescue her. As he was paddling her away, the raven appeared and caused a great storm.

A typical story line in mythology goes as follows: The father sees the error of his ways and saves his daughter; the evil suitor attempts to take her back; the father vanquishes the suitor. In the Inuit myth, however, things go slightly differently.

The father, fearing for his own life, throws his daughter overboard into the storm and back to the raven. The girl begins to sink. She grabs the side of the boat to hang on to. The father takes out his knife and cuts off her fingers to keep her from climbing aboard. The girl sinks and becomes the goddess of the sea. Her fingers and thumbs become the seals and whales of the ocean. She is angry much of the time—understandably so—and causes storms to thwart the hunters. But she is soothed when a shaman swims to the bottom of the ocean and brushes her hair (having no fingers, she can't hold a brush), and then she relents and lets hunters safely venture again. I hope Sedna is happier now, at the bottom of the ocean, and, especially, up in the sky, than she was with her creepy father or raven husband.

Contemporary Inuits make fantastical carvings of their mythological figures. The weekend before the press conference at which I was going to reveal Sedna to an unsuspecting world, I signed on to eBay and found that Sedna carvings could be had for a few hundred to a few thousand dollars. To celebrate the discovery, I bought what to me was a particularly nice—and particularly affordable—carving in which Sedna has the body of a seal, the arms of a woman, hands with no fingers, and a mermaidlike face. The Sedna carving sits in the center of my desk to this day, surrounded by other mementos of planetary discovery. The eBay bidding on the Sedna carving closed on Sunday night. The press conference was on Monday. By the end of the day on Monday, I checked and saw that prices of Sedna carvings had gone up by a factor of two. Yes! Maybe I had a future in Wall Street insider trading when discoveries in the solar system finally came to an end.

The name was a hit. I was surprised to discover that a good name with an interesting story behind it could lead people to have an emotional connection with an unseen object in space, though perhaps it shouldn't have been such a revelation to me, given people's attachment to Pluto. Quaoar never really caught on, but Sedna struck a nerve. Newspaper headlines proclaimed, "Welcome Sedna!" My mailbox began to be flooded with drawings from schoolkids who crayoned in nice red Sedna in the solar system right after Pluto. Astrologers quickly hit on the story of Sedna to declare that Sedna would herald a new feminine influence over environmental stewardship. Or awareness of child abuse. Though none of the astrologers agreed with one another, they certainly found the name and the story compelling.

The only problem with the name was that I had jumped the gun and broken the astronomical naming rules.

This was not the first time I had broken the rules. When I announced the discovery and name of Quaoar, it turned out that I had not sought approval through the proper channels in the International Astronomical Union. I didn't realize that I was supposed to have tracked down the Committee on Small Body Nomenclature of the International Astronomical Union and proposed the name, allowing the august committee to deliberate and declare whether or not my name was appropriate. Luckily, the name Quaoar was perfectly appropriate, so the CSBN of the IAU promptly approved the name without my having gone through the channels, though eventually it did make me fill out the official form.

No harm done, and it seemed to me that nobody cared much. At least, that's what I thought.

Unknown to me there was a group who cared a lot. Somewhere in the far corner of the Internet was a chat group composed of astronomy enthusiasts who had appointed themselves the celestial police. I didn't know they existed until one day a student of mine pointed me to their chat site with the comment "Wow, they really hate you, don't they?" And it did seem as if they hated me, or at least felt that antagonistic indignation that can be pulled off particularly well on the Internet.

They were angry because with Sedna, I had not only broken the rules, I had done so on purpose. At the time of the announcement of Sedna's existence, we didn't quite have enough data for Sedna to officially qualify for a name—it would take us another few months to have what we needed. The rules on when an object qualifies for a name are obscure, uninteresting, and designed to keep names from being given to insignificant asteroids that are seen a few times, then never again. Nonetheless, they are the rules, and to the zealous enthusiasts, they must be followed at all cost to prevent astronomical chaos from breaking out.

I admit that in the week before the announcement, even I worried a bit about breaking the rules. I am, by nature, a rule follower. But I really wanted Dutch to be Sedna in time for the announcement. I thought it mattered—and, it turned out, based on those crayon drawings, it did. Finally I decided I would buck the rules, though politely. I called Brian Marsden, an astronomer at Harvard University who was, in my opinion, the gatekeeper of the solar system. He was the person to whom you sent the very first announcements of discoveries. He checked that your calculations were right. He put your discovery on the official list.

And he was always the first to be amazed and say, "*Wow!* What a great discovery." Brian was also the secretary of the Committee on Small Body Nomenclature. I told him what I was planning to do. He asked if he could tell the chair of the committee ahead of time. Of course, I said. Everyone agreed that a name was a good thing and that Sedna was a good name.

To the chat group, though, I was a rule breaker in need of punishment. One particularly agitated enthusiast tried very hard to prevent me from officially naming Sedna Sedna. Before Sedna was quite eligible for an official name, he proposed, through the official channels, that an unremarkable, hitherto anonymous asteroid—which was nonetheless eligible for a name—be named Sedna, after the Inuit goddess of the sea. No two things in the solar system can have the same name, so my Sedna would have had to get a different name.

"Rejected," declared Brian Marsden. Names of important mythological figures would be used only for important astronomical objects.

The enthusiast next proposed to name the unremarkable asteroid after Kathy Sedna, a Canadian singer.

"Clever," responded Brian Marsden, who, being in charge also of when things are eligible for names, quickly realized that my Sedna was now eligible and made sure the name became official.

I found all of this pretty amusing at the time. It was proof to me that names do, in fact, matter, and I even found it moving that there were people who cared so much about the details of scientific naming. I didn't know that in just eighteen months some of these very people would have a hand in almost stealing the most important discovery I had ever made.

Sedna remained Sedna. And with all of the crayon drawings showing Sedna's rightful place in the solar system, Sedna was surely a planet, right? It's true that I had argued against Quaoar and Pluto being planets on the basis of their being in the middle of swarms of similar objects. To me, it made no sense to pull one or even a few objects out of the swarm and call them something other than part of the swarm. But Sedna was, as far as we knew, all by itself. There was no swarm of objects out in the region of space where nothing was supposed to be found. Couldn't it be called a planet? That, too, made no sense. Sedna would eventually be found to be part of a swarm, too. If we called Sedna a planet now, when that swarm was finally found, we would have to go through the process of planetary argument all over again. It seemed better to put Sedna in the right place to begin with.

Besides, Sedna was smaller than Pluto. In the beginning, we had been certain that Sedna would be bigger than Pluto. It was so bright! But when we finally got a chance to look at it with the Hubble Space Telescope, thinking we would get to see a little disk of a planet, all we saw was a tiny point of light, and that tiny point of light told us that Sedna was no more than about three-quarters the size of Pluto. How could that be? The answer is always the same: albedo. Sedna has an even more reflective surface than Quaoar, so part of the reason it is so bright is simply that reflectivity. Still, three-quarters the size of Pluto is big! No one else alive had ever found anything bigger in the solar system. But finding bright things that are almost certainly bigger than Pluto only to realize that, well, no, they aren't actually bigger after all, gets old.

I hadn't thought about it for a while, but it had been four years since my bet that someone would find something bright enough to be called a planet within five years. Much had happened since that bet. We had found Quaoar, at half the size of Pluto; Sedna out where the solar system was supposed to end; and dozens of other smaller objects that were nonetheless among the biggest things anyone else had ever seen. But we hadn't found anything yet that would qualify for the bet to be won.

We announced the discovery of Sedna in February 2004. My bet ended on December 31. I had a little more than ten months to find something truly large, or I was going to lose.

I hate to lose.

And even worse than losing, I hate being stupid.

One thing nagged at me. I had almost missed Sedna. Sedna is so far away and therefore moves so slowly that the computer program I had written had almost ignored it. If Sedna had been just a little farther away and therefore moving just a little more slowly, we would never have found it. My computer program would have declared it to be a stationary star and kept searching. If Sedna was there and had been almost ignored, couldn't there be something far out there that *had* been ignored? Finding such distant things would be crucial for testing my hypothesis about the birth of the sun and the odd population of distant objects that would have been created. But also, if we can see things that far away, they have to be big. It occurred to me that one of the best places to look for planets might not be in the remaining unexplored parts of the sky but in the many, many pictures I had already taken. If there was a planet already there that I had missed the first time around, I would, indeed, feel I had been stupid. But as I had learned earlier, the trick was not to figure out how not to be stupid, the trick was to be smart instead.

I spent most of that summer in my office slouched in front of my computer screen, writing, testing, and rewriting software. About halfway through the summer, one of the other professors on my hallway started commenting.

"You never move," he said.

"My fingers move."

In fact, my fingers moved a lot. I had thoroughly rewritten all of the computer software. Chad had written the first version without the benefit of having any of the data at the time. With the luxury of data, I could rewrite it to work better, run faster, search farther, and see fainter objects. I was ready. I started spending my days not just looking at the new pictures coming from the telescope the night before but also scanning the thousands of pictures that I had stored on the disk drives of my computer.

Someone watching over my shoulder that summer would have seen an incredibly monotonous sight: Mike presses a button; a new series of images begins blinking on his screen; he stares for three seconds; he presses a button marked "no"; new images appear.

I did this for hours a day. My posture got even worse. My back ached. But I was discovering things in the old pictures. The first time around, we had missed a lot. This time, I didn't want to miss anything.

I think of this period in the fall of 2004 as one of the most fertile in my life. Still, though, there were no planets, and I was losing my bet. I was working longer hours, sleeping less, all in the hope of getting through all of the data before the end of the year. I really did not want to lose the bet. If there was something to be found in the old pictures, there was nothing, absolutely nothing, that would stop me from finding it. Well, almost nothing.

At the beginning of December, taking a rare break from looking at my old pictures, someone else showed me a picture of something I had never seen before. The moment I saw it, my mind flashed back to images I remembered having seen in high school. In 1982, a Russian Venera spacecraft sent back the first—and still only—color pictures from the surface of Venus. Venus is a tough place to take pictures from. The surface has an atmospheric pressure ninety times higher than the earth's and a temperature of more than eight hundred degrees, which would melt the lens of any camera. The Russians therefore built the camera inside a giant can to keep the extreme pressures and heat out as long as they could. To see Venus, a periscope popped out of the can and scanned around. Even so, the whole contraption lasted only two hours before it died.

The pictures that the Russians sent back from Venus have a peculiar characteristic to them. Because of the periscope, they are oddly distorted, as if they had been taken by a fish-eye lens. Because of the thick clouds of sulfuric acid that cover Venus, among other things, the color pictures have an oddish orange glow and appear almost to be black and white. They are hard to mistake for almost anything else.

I had been spending most of my time those past few months staring at a huge computer screen hoping to be the first person ever to see a new big thing moving through the distant regions of space. That morning, I stared at a smaller screen and examined a black-and-white image with an orange tint to it and an oddly distorted view like that through a fish-eye lens. It wasn't Venus. In the middle of the oddly distorted view was a little bean-sized object. Looking at the sonogram, Diane and I, along with our doctor, were the first people to see the tiny movements of a little heart beating.

"Hey!" I said. "It looks like the Venera lander pictures of the surface of Venus."

"You're insane," Diane said.

We told our families on New Year's Eve. Mine were visiting from Alabama. Diane's lived in town. Everyone sat down to dinner.

I began: "Before dinner, I'd like to make an announcement."

I had been saying this at every family dinner since Diane and

I had been married. I usually then proceeded to say, "It's time to eat." People who are regulars at our dinners barely look up while awaiting the now-tedious punch line.

My family, however, had never heard the joke. They gasped slightly. Diane's father quickly interjected, "He says this every time, just ignore him."

Everyone calmed down and ignored me, until I said, "We're expecting a baby girl in July. Her real name will come later, but her current code name is Petunia."

That night, as the clock struck twelve, my five-year bet came to an end. I lost the bet, but I didn't feel so bad. Instead of seeing the end of the solar system, I saw that everything was just beginning.

Chapter Seven

RAINING = POURING

The next morning, January 1, 2005, my whole household woke up early to walk down to the Rose Parade, which winds its way through Pasadena every New Year's Day. In the still-dark early morning I was awake in time to find Jupiter bright in the sky before the sun came up. Jupiter, Saturn, Uranus, Neptune, Pluto. That was it: the end of the planets.

Maybe.

Unbeknownst to anyone—well, except for Diane, to whom I told everything, and my parents, who were visiting, and all of my students, and a few friends here and there—two days after Christmas I had discovered the brightest thing I had yet seen. I didn't know for sure how big it was, so I was not in time to win my bet, but something that bright might well be a planet. In honor of the season when it was discovered, I called it Santa.

A few years earlier, my first reaction to the discovery of Santa would have been: I bet it's bigger than Pluto! I've finally found the tenth planet! By now, though, I was a bit more skeptical.

Quaoar and Sedna had both fooled me with their anomalously frosty surfaces, which made them appear much brighter than I expected. But even if Santa's surface was as anomalously frosty as Sedna's, it would still mean that Santa was the size of Pluto. But what if Santa were even frostier? What if Santa was covered in, say, pure ice, which would make it even shinier and brighter than Sedna? I wasn't going to get my hopes up too much.

I sent e-mails to Chad and David telling them what we had found. I was careful not to definitively declare the discovery as bigger than Pluto, but I did mention that if it had a dark surface—as we had long assumed most objects in the Kuiper belt did—it would have to be almost as big as Mercury.

Over the next week, Chad, David, and I raced to see who could find old pictures of Santa to figure out what kind of orbit around the sun it had. Chad won and declared the orbit thoroughly normal. "Normal" in the case of the Kuiper belt means elliptical and tilted, but still within the swarm of all of the other Kuiper belt objects. After the oddness of Sedna, this normal orbit was almost a relief. At least *something* about the Kuiper belt was making sense.

Today I know Santa by its official name, chosen by David: Haumea. The mythological Haumea is the Hawaiian goddess of childbirth. Her many children, which compose a large subset of the population of Hawaiian deities, were broken off from different parts of her body. The astronomical Haumea has been equally prolific. In the years since its discovery, we have found many other objects in the outer solar system that we can now trace back to having originally been part of the surface of this object. We think that at one moment early in the history of the solar system, a much larger Haumea was smashed by another icy object in the Kuiper belt traveling at something like ten thousand miles per hour. Luckily for Haumea and for astronomers today, the impact was only a glancing blow. Had it been more head-on, Haumea would have thoroughly shattered and dispersed to the ends of the solar system. Instead, the glancing blow left the center of Haumea mostly intact, but large chunks of the surface went flying into space, while Haumea itself was left spinning faster than almost anything else in the solar system. Some of the chunks that were blasted off the surface didn't go far; at least two are now in orbit around Haumea as small moons (when we first discovered these we called them Rudolph and Blitzen, but now they are named after children of Haumea: Hi'iaka, the patron of the Big Island of Hawaii and the goddess of hula, and Namaka, a sea spirit). Many more of the chunks were blasted so hard that they escaped Haumea entirely and now form a virtual cloud in orbit around the sun.

It also turned out that I was right not to get my hopes up about the size of Santa/Haumea. We learned that Haumea *is* covered in pure ice, and it *is* smaller than Pluto.

None of this was obvious when Santa/Haumea was first discovered. It just looked like a normal, albeit extra-bright, object in the Kuiper belt. David was the first to notice something strange: It got brighter and fainter every two hours, a fact that he quickly surmised was due to the fact that Haumea was oblong and tumbling end over end every four hours.

Huh, we all said.

Next we discovered two moons.

Weird, we all thought.

It wasn't until eighteen months after the discovery that the final pieces of the puzzle came together. It was around midnight at a beach hotel on the island of Sicily. Kris Barkume, another graduate student of mine, was going to give a presentation the next morning at an international conference on the subject of her Ph.D. thesis, which was a study of the many moderately bright objects that had been discovered by Chad, David, and me. One subset of these objects appeared unusually icy compared to everything else out there. I had asked her to concentrate on trying to understand what might be going on with those objects. By the midnight before her talk she had learned much, but she still didn't really have an explanation. We sat down on the sofa in the lobby of the hotel so that she could go over her talk with me.

We kept looking at the data on the odd icy objects, and still no obvious explanation came to mind. Finally she said, "Oh, and you know what's funny? Their orbits around the sun are almost identical."

They are?

"Yeah, look. And you know what else is funny? Santa has almost the same orbit."

In my scientific life, most of the discoveries come as the result of seeing something for the first time. A picture appears on my screen and I suddenly know something big is out there. I know no one has ever seen it before, and I feel that little charge. This time it was different. There was no obvious picture on the screen. We were just sitting on the sofa. But instead of a little charge, I felt a full jolt of instant understanding. It all suddenly made sense. Santa's spin, Santa's moons, the little icy objects flying around it: They were all caused by that one glancing blow millennia ago; the moons and the strange little icy pieces flying around were all the debris blasted off the surface in what we now know to be the largest impact in the outer part of the solar system. Ah *ha*!

Kris gave her talk the next day, skillfully laying out all of the pieces of the puzzle that we had just discussed the night before and reassembling them to tell the story of one of the most dramatic events in the known history of the outer solar system. Everyone gasped.

It took us years after the initial discovery of Haumea to find out all of these details. Even today we're still studying Haumea and learning more and more. In the days following the discovery, back when Haumea was just Santa, I knew little more than that there was a big bright object out there waiting for me to study it in detail at the start of the year.

In addition to studying Santa, I had other things on my mind that New Year. Though I had pushed hard to finish looking at all of the old pictures to find really distant objects before the end of the year, I had not only run out of time, I'd run into distraction. I admit that I spent less time thinking about the science of the outer solar system than I did worrying about the science of embryonic growth and early childhood development. Hours that could have been spent staring at pictures of the night sky were spent, instead, reading about statistics of timing of childbirth and first smiles. I was still obsessed; I had just changed the main object of my obsession.

· · ·

I had been at work on January 5 for only a few hours when I decided to get up and take a walk. I needed to walk down the street and get some lunch. I had some things to think about. Lunch that day was the same as lunch most every day. I went to the same busy corner just down the road from my office; I ordered the same sandwich from the bagel store; I sat staring into the same steaming cup from the coffee shop next door. I like things that stay the same. The sun was shining and the seats on the outside patio were packed and everyone was emerging for a several-day break in the record-setting rains that were pummeling southern California that winter. From my spot on the patio I could see the temporarily snow-covered peaks of the San Gabriel Mountains just a few miles to the north. To me, there is almost nothing more relaxing and serene than this particular cup of coffee drunk at this particular spot on the planet Earth at this particular moment in the year, when the winter storms have come from across the Pacific Ocean and cleared the skies and coated the mountains, and the sun, low even at high noon in the clear skies just a few days after the winter solstice, is shining on the tables outside and quickly melting the snow on the mountains beyond.

I particularly like the stability and predictability of this spot when I know that everything is about to change. I sat in this same spot staring at these same mountains in the last hour before my wedding, thinking about the future, thinking about the past, suddenly remembering that I had left my bow tie at home. It was the same spot where I sat with Diane for hours on a workday and realized that she was choosing to stay and sit with me rather than going back to work and that I had been stupid all along. Later, I sat in the same spot with Antonin Bouchez as he convinced me not to quit searching the skies. And, though I didn't know it at the time, six months from now I would momentarily pause at this same spot—no time for sitting now!—as the last stop as I was taking Diane to the hospital for the birth of our Petunia, thinking only about the impending present and how long the night ahead was going to be.

This clear January day, one in which I watched the waterlogged people enjoy the fleeting sun and stared at the snow quickly melting on the mountains, was a day I would remember as well as those other momentous days at this spot. After sitting on the patio, drinking my coffee, and staring one last time at the mountains, I walked back to my office, sat down at my desk, and carefully composed a short e-mail that I knew would set in motion a series of events that would lead to a change in our view of the solar system. Eventually the news would spread across the planet, but, for now, I sent copies to only two people: Chad, 2,500 miles west of me on the Big Island of Hawaii, and David, 2,500 miles east of me at Yale University. They were about to become just the third and fourth people in history to know what I had known for several hours (Diane was, of course, the second) and had been thinking about as I stared at the mountains over lunch: The solar system no longer had nine planets.

When I had left home to go to work that morning, the nine-planet solar system was still intact. Sure, the discovery of Santa was exciting, but given our track record of discovering things that turned out to be smaller than Pluto, I was pretty sure that Santa would be, too. It seemed likely that the solar system would retain nine planets. It seemed likely until I sat down at my desk that morning and discovered the tenth. There it was, moving across the sky, visible on a series of pictures blinking across my computer screen. Two weeks after the discovery of Santa, the almost-planet, I had found the real thing.

There aren't many chances in life to write an e-mail like the one I sent to Chad and David. I'd thought all through lunch about how exactly I would word it. I went for carefully calculated obscurity:

Subject: why we get up in the morning

And then I went on, staccato style:

new bright object
please sit down and take a deep breath

> mag = 18.8, making it brighter than anything out there ex-
> cept Santa
> distance = 120 AU
> and, by the way, if you moved Pluto to 120 AU it would be
> about mag 19.7

That's all they needed to know to understand that the solar system was, from that day on, a different place. To most people, all of this would be more or less nonsense (at least I hoped, in case there were prying eyes; I was, I thought, overly paranoid, but in the end it turned out I was not nearly paranoid enough), yet Chad and David would instantly see the significance of each of the lines.

> new bright object

We had just discovered Santa two weeks earlier, and I was sure they would assume that was the object I was referring to. What else would I be writing about? No one expects the next one to come so fast.

> please sit down and take a breath

Okay, I have a melodramatic streak.

> mag = 18.8, making it brighter than anything out there ex-
> cept Santa

Astronomers describe the brightness of their objects in "magnitudes," and "mag = 18.8" immediately told Chad and David that the new object was bright, at least for something out there in the region of Pluto. But this was only the second-brightest object we had found to date, and it wasn't even as bright as Pluto. The next line was designed to get them to fall out of the seats that I had previously asked them to sit in.

distance = 120 AU

The phrase "120 AU" means 120 times the distance from the sun to the earth, or about 12 billion miles. Even to astronomers, the phrase "12 billion miles" generally means nothing other than "really far away." But 120 times the distance from the sun to the earth is packed with meaning. It is farther than anything that had ever been discovered in orbit around the sun, and almost four times more distant than Pluto. Finding something at this distance was a major discovery, regardless of what it was. But something so far away would be expected to be so faint that it would be just barely visible in our telescope. This object was not just barely visible, it was almost the brightest thing we had ever discovered. The brightness ("mag = 18.8") combined with the distance ("distance = 120 AU") meant that I was writing about something that must be larger than anything we had found in all of the previous years of our searching. The next line of the e-mail drove the point home, in a feigned attempt at nonchalance:

and, by the way, if you moved Pluto to 120 AU it would be about mag 19.7

Pluto is much closer to us and to the sun than this newly discovered object, so it appears to be much brighter; but if you moved Pluto out to the same distance as the new object it would be almost three times fainter than the new object (which, in astronomers' archaic system, would mean that it had a higher magnitude). If you have two objects at the same distance from the sun and one is brighter than the other, chances are that the brighter one is bigger than the fainter one. Chances were that the newly discovered object was bigger than Pluto. Chances were that the nine-planet solar system had just come to an abrupt end on that early January morning.

I pressed the "send" button on the e-mail and sat back to think about the significance. Nothing this large had been found in the solar system in more than 150 years; no person alive today had ever found a planet; history books, textbooks, children's books would all have to be rewritten. But I don't remember thinking any of those things. All I can remember thinking is that we were only five days into the New Year and Diane and I had, just a week before, told our parents and friends that we were expecting our first child; a week before that I had discovered Santa, which would eventually spawn the biggest astronomical controversy in years; and now I had found something bigger than Pluto.

Wow, I thought, this sure is going to be a busy year.

. . .

I sent one more e-mail that afternoon before diving in to learn what I could about the new object. It was to Sabine, the friend with whom I had made the bet five years earlier.

> Would you be willing to grant me a five-day extension on our bet?

She said that she would.

By the end of the week, David had tracked the object down on some recent pictures he had taken, and Chad had followed it into the past for decades. We knew the orbit precisely. The orbit was, like that of Santa, relatively normal. It was scattered. It was so far from the sun right now only because we had caught it at its most distant point. It's on its way back in but will take a while to get there. The object takes 557 years to go around a full orbit, so it will be half of that—278 years—before it is at its closest point to the sun. When that happens, it will be closer than Pluto and thus, presumably, brighter as seen from the earth. I can't wait to see it.

Clyde Tombaugh found Pluto in 1930 but spent much of the following decades searching for whatever else might be out there in the distant regions beyond Neptune. He never found anything else. As the story is usually told, this was because he was using the old technology of photographic plates, which were simply not good enough to see what we now know as the myriads of objects out there. But we now know that the story is not quite so simple. If Tombaugh had been looking 278 years earlier or 278 years later, our new object would have been as bright as Pluto, and he would have found both.

It's interesting to ponder what people would have thought in the 1930s if not just Pluto but also this new object had been found. Both are on crazy elongated orbits. Both appear significantly smaller than the giant planets. And their orbits cross. I'm pretty sure that the similarities to the asteroid belt discoveries 130 years earlier would have led everyone to conclude that these were simply the biggest members of what would turn out to be a huge population of similar objects. And they would have been right. Instead, though, the solar system was arranged such that at the time of the development of large photographic plates and the first major survey of the outer solar system, only Pluto was close enough to be seen. Too bad for us not to have been provided with such obvious clues to the nature of the outer solar system. But good for Pluto, since it got to be everybody's favorite oddball planet for more than seventy-five years. Though it wouldn't be for much longer.

We didn't refer to our discovery as "this new object" for long. We quickly gave it a code name. Unlike Flying Dutchman or Santa, which were inspired by circumstances of the discovery, we had had a name waiting for this one for a long time. Since the earliest days of surveying the outer solar system with the photographic plates, I had always had a well-considered code name for the hypothetical object bigger than Pluto. In coming up with the name, I had thought that it was best to keep the *X* for the apocryphal Planet X beyond Neptune. And I had thought that Venus shouldn't remain the only female among the planets. And finally, I thought that the name should be mythological.

With those criteria, I was left, as far as I could tell, with only one choice. We called the new object Xena, after the eponymous heroine of *Xena: Warrior Princess,* the campy, female-empowered television take on Greek mythology starring Lucy Lawless. It was true that the name Xena was only TV mythology instead of real mythology, but as I liked to point out for the next eighteen months, as the name got more and more widely known, wasn't Pluto named after a Disney dog? Whenever I made that joke publicly, about half the people in the room actually thought I was serious.

A few weeks after the discovery of Xena, Chad got a chance to swing the giant Gemini telescope, at the summit of Mauna Kea on the Big Island of Hawaii, in its direction. Chad now worked at that telescope, so getting a little time at the discretion of the director to look at something that was clearly bigger than Pluto was no hard task. When he looked at Xena's surface, we had our first confirmation that Xena was something special. Xena looked like Pluto.

By "looked like Pluto" what I really mean, to be more precise, is that the sunlight bouncing off Xena contained within it the unmistakable signature of a surface covered in solid frozen methane. Nothing else in the Kuiper belt looked like this, with one exception: Pluto.

It was one thing to make the quick calculation to know that Xena was bigger than Pluto. But we had been looking at new Kuiper belt objects for a long time, and we had never seen anything that looked like Pluto.

I went home that night and told Diane about the methane.

"So it's a planet?" she said.

"No," I quickly pointed out. "It means Pluto is not a planet."

"But if there are only two of them out there that look like this, and they both look different from everything else, why not just call them both planets?"

I went over my usual litany: Pluto was simply the largest— now the second largest!—member of a huge population in the Kuiper belt. Singling it out for special planetary status really made no sense at all.

"Okay, but think about your daughter."

Huh?

"Having her father discover a planet might mean that someday she'll be able to afford college."

Diane was joking. At least mostly.

I remained adamant.

She pressed: "Didn't you used to joke that your definition of a planet was 'Pluto is not a planet, but anything that I find that's bigger is'?"

Yes. I had made that joke. But it was a joke.

Diane was in her energetic superwoman second trimester. If I was working late trying to figure out something about Xena or Santa, she would stay awake even later looking at baby magazines. If I woke up early to try to look at a few pictures of the sky just as they were coming off the telescope, she would already be awake looking at pregnancy books. As long as she was fed, she was unstoppable.

"But really, you are going to have people arguing that it *is* a planet, and you're going to stand up and say, 'No no no no'? If Petunia is a really cute child, will you go around and point out that, really, she is not so cute because, well, her nose is a bit big?"

Well, only if it *is* big.

"Don't you think it would overall be better for astronomy to have new planets discovered rather than have old planets killed?"

I think it's better to get it right.

"But don't you think the public would be more excited and engaged in astronomy and in science if there were new planets being discovered?"

Enough, woman who needs no sleep! It is past my bedtime! But I will ponder your suggestions in the morning when I wake up, which will be long after you've already risen.

As winter waned to spring, three independent trains of thought ran through my mind. If I ever got to spend more than a few hours in a row thinking about one of them, I would suddenly sit up with a start and remember one of the others and start thinking about that one, before I suddenly remembered the third, and then the process would start all over again.

The first ticking calendar was strictly biological. Petunia was getting bigger. Her bones were hardening. Her eyebrows were growing. She had a July 11 due date, and though there was not much I could do to influence anything, I could nonetheless obsess about what, precisely, a due date means. I asked anyone who I thought might have some insight. I know, for example, that due dates are simply calculated by adding forty weeks to the start of the mother's last menstrual cycle. But how effective is that? How many babies are born on their due dates?

few days of their due dates) or short and fat (if there is quite a wide range around the due dates). One thing I know, though, is that the bell would have a dent on the right side. At least around here, no kids are born more than a week or two after their due dates. Everyone is induced by then.

I am usually capable of allowing myself to give up on trying to get the world to see things in my scientific, statistical, mathematical way. But this mattered to me. If I was at a dinner party with Diane and the subject of due dates was broached, Diane would turn to me with a slightly mortified look in her eyes and whisper, "Please?" I would rant about doctors. About teachers. About lack of curiosity and dearth of scientific insight and fear of math. I would speculate on the bell curve and about how fat or skinny it would be and how much it might be modified by inductions and C-sections, and whether different hospitals had different distributions. Inevitably the people at the dinner party would be friends from Caltech. Most had kids. Most of the fathers were scientists. Most of the mothers were not. (Even today things remain frighteningly skewed, though interestingly, most of my graduate students in recent years have been female. Times have no choice but to change.) As soon as I started my rant, the fathers would all join in: "Yeah! I could never get that question answered, either," and they would bring up obscure statistical points of their own. The mothers would all roll their eyes, lean in toward Diane, and whisper, "I am *so* sorry. I know just how you feel," and inquire as to how she was feeling and sleeping and how Petunia kicked and squirmed. (As an aside, my female graduate students wanted to know the answer to my question, too, and were prepared to rant alongside me. Times have no choice but to change.)

There was another calendar ticking, too. At the moment, the sun was almost directly between the earth and Xena. We knew

Our child-birthing class teacher: "Oh, only five percent of babies are actually born on their due dates."

Me: "So are half born before, half after?"

Teacher: "Oh, you can't know when the baby is going to come."

Me: "I get it. I just want to know the statistics."

Teacher: "The baby will come when it is ready."

I asked an obstetrician.

Doctor: "The due date is just an estimate. There is no way of knowing when the baby will come."

Me: "But of your patients, what fraction delivers before, and what fraction delivers after the due date?"

Doctor: "I try not to think of it that way."

I propose a simple experiment for anyone who works in the field of childbirth. Here's all you have to do. Spend a month in a hospital. Every time a child is born, ask the mother what the original due date was. Determine how many days early or late each child is. Plot these dates on a piece of graph paper. Draw a straight line for the bottom horizontal axis. Label the middle of the axis zero. Each grid point to the left is then the number of days early. Each grid point to the right is the number of days late. Count how many children were born on their precise due dates. Count up that number of points on the vertical axis of your graph and mark the spot at zero. Do the same with the number of children born one day late. Two days late. Three. Four. Keep going. Now do the early kids. When you have finished plotting all of the due dates, label the top of the plot "The distribution of baby delivery dates compared to their due date." Make a copy. Send it to me in the mail. My guess is that you will have something that looks like a standard bell curve. I would hope that the bell would be more or less centered at zero. It would either be tall and skinny (if most kids are born within a

Xena was out there, but we couldn't see it. When Chad had looked at the surface of Xena and realized it looked similar to Pluto, he had done so at almost the last possible moment. Xena had been low in the western sky just at sunset. A few weeks later Xena set with the sun, and we couldn't see it anymore. But slowly, the earth was moving around the sun, and Xena was eventually going to reappear on the other side, this time in the early-morning sky. As desperate as we were to learn more about Xena, we had no choice but to wait. Our first chance to get a good look would not be until September. I made sure that we were scheduled to be at the Keck telescope then. Other than that, there was little we could do except try to keep from telling people. I was looking forward to springing my news on an unsuspecting world.

But another calendar was ticking, too, this one involving two different moons. One of the people I had told about Santa was Antonin Bouchez, the former graduate student of mine who had convinced me not to quit two years earlier. He now worked at the Keck telescope helping to develop a fancy new technologically intensive way to make extra-sharp pictures with the telescope. Usually, when you take a picture of a star or planet or anything else with a telescope, the earth's atmosphere blurs that object a little bit, preventing you from seeing the smallest details. This blurring was the reason that we were never certain at first how big the things we kept finding were; their little disks were blurred out enough that they all looked the same. Antonin had been hired to work on a project to fix this problem. The trick is to take a powerful laser and shoot it out the front of the telescope into space. The light from this laser traverses the earth's atmosphere, and then, right when it is about to shoot out the top, it encounters a thin layer of gas that has been burned off from asteroids as they enter the earth's atmosphere. The laser light is precisely tuned to bounce off the asteroid gas and return back to the earth. If you then take your telescope and point it right at the location of the laser, you see a little spot of light—an artificial star!—in the sky. The real trick then happens. You take that picture of the laser beam, which has been distorted by the earth's atmosphere, and you bounce it off a fun-house mirror that is warped precisely so that the picture of the laser is as sharp as you know it is supposed to be. And then you do it all over again one-hundredth of a second later, using a different fun-house mirror shape, as the roiling of the earth's atmosphere distorts the laser beam differently. If you can get all of the laser light going in the right place and the computers calculating fast enough and the fun-house mirrors warping accurately at your command, you can then take a nice long picture of the sky, and you will see a beautiful pinpoint of a laser beam, just as you shot it. That would be a lot of work just to see a laser beam; but if you pointed the laser beam directly at something else in the sky that you really cared to look at, you would also be perfectly correcting the light coming from *that* object, too.

One of the very first times that Antonin and the rest of the team hard at work at the Keck Observatory got all of the pieces in place and pointed up to the sky to test things out, they looked at Santa—and they found that Santa has a moon. We named it Rudolph.

Discovery of moons is extremely helpful, since it means that you can weigh the object using simple high school physics. Once you know how far away the moon is from the object and how fast it goes around the object, you've got all the information you need. Now that we knew that Santa had a moon, we only had to observe it a few more times and we would know how fast it was moving and how far away it was.

For me, all this meant was a lot of waiting. The first ticking clock was, strangely enough, that of our own moon. Because the laser system that Antonin was working on was still experimental, no one wanted to potentially waste the most valuable observing time. The laser was allowed to experiment on the telescope only when the moon was full—when it was bright time and many of the astronomer's favorite targets in the sky were washed out by the glare of the moon. So though we knew about the moon of Santa quickly, we didn't get our second look until twenty-nine days later, when the earth's moon came around full again.

After twenty-nine days Rudolph had moved, but we now had no idea if it had gone around several times and come back or if this was still its first revolution. We had to wait twenty-nine more days to get our next look. At this point it was close to where we had seen it the first time. This would all make sense if the moon took about fifty days to go around Santa, but I couldn't be certain for twenty-nine more days. The fourth time we saw it, we knew for sure: Rudolph goes around Santa every forty-nine days. It took one final measurement for confirmation to make sure we had the distance down, too. Combining the time period of the orbit with the distance from Rudolph to Santa allowed us to know that Santa weighs only one-third as much as Pluto. It was a relief, almost. We had finally been careful not to get our hopes up too much.

Just because there was Santa's moon to track and due dates to fret about and Xena to anticipate, it didn't mean that new pictures weren't rolling in every night.

The universe speaks to me in strange ways. One day when I was in graduate school, two of my best friends who didn't know each other separately told me that they were each expecting their first child. Strange, I thought, when the second friend told me. What could this mean? What is the universe telling me? I thought about it for a while and came to the conclusion that, clearly, this meant that my sister, who had been married for several years by this point, must be pregnant. What else could the universe be trying to say? I talked to my mother that very night.

"I think Cammy is pregnant."

"What?" my mother replied. "Have you just talked to her?"

"No, the universe told me."

My mother never quite knows how seriously to take me.

When my sister called my mother the very next day and said, "Mom, guess what?" my mother said, "You're pregnant." My sister was extra flabbergasted when the answer to "How did you know that?" was "Mike told me."

When my sister called me, she asked, "Are you studying astronomy or astrology?"

Apparently my communication with the universe is not always so reliable—I just missed the signs this time. Petunia was growing. Rudolph was revolving around Santa. Xena—mighty Xena!—was swinging toward the night sky. The universe was presumably trying to warn me what I didn't learn until April 3. While sorting through pictures from a few nights earlier I saw the brightest thing on my screen ever. Brighter than Xena. Brighter than Santa.

"Here we go again," I thought.

The subject line of the e-mail that I sent to Chad and Dave was:

raining = pouring

It was raining so hard that I was in danger of drowning.

We named this new object, found two days after Easter, Easterbunny.

Easterbunny was so bright that it was probably bigger than Pluto, too, or maybe the same size. We quickly took a look at it

with the Keck telescope and realized that, like Xena, Easterbunny had a surface that looked like Pluto's. The solar system
had gone from one to two to three Pluto-like objects in just three
months.

I almost felt bad. This was too much! How was I going to
give everything the attention that it deserved? I needed a plan,
now.

Our goal was to follow good scientific practice and announce
the existence of these objects to the world with a full scientific
account in a scientific journal. But full scientific accounts take
time. We had done well with our previous discoveries. Quaoar
had taken about four months from discovery to scientific paper.
Sedna had taken about the same. We were pretty proud of our
speed. But even if we could keep up the fast rate, we suddenly
had Santa and Xena and now even Easterbunny to write papers
about.

David and Chad and I made a plan. Santa had been discovered first, and we knew the most about it already. We would each
write papers on different aspects of it. Whenever the first paper
was finished, we would have a low-fanfare announcement. We
knew that Santa was smaller than Pluto, and we didn't yet know
all of the details of Santa's massive collision and debris field, so
we thought there would not be too much interest in it for the
public. My goal was to get a paper on Santa finished before the
birth of Petunia, since I still had a little free time. Her due date
was now only three months away.

We would then save the big excitement for Xena and Easterbunny, which were sure to cause a stir. We were scheduled to be
at the Keck telescope in September to get a first really good look
at Xena. With some intense work we could have a scientific
paper ready a month after that (the delusions of first-time parents astound me to this day) and make the announcement

around the beginning of October. I liked this back-to-school timing, as I thought having the announcement of one or two new objects bigger than Pluto would be the sort of thing that schoolkids would think was cool to talk about in class.

The plan required writing perhaps the three most important scientific papers of my life in under six months while having my first child. No problem, I thought.

Diane was having that last spurt of energy that comes in the weeks before delivery. The spare bedroom, which for years I had dubbed the "bike and computer room," was suddenly transformed, with a crib and pale green walls and a collection of infant clothes waiting for an owner. I acquired a bit of sympathetic energy and redoubled my writing efforts so I, too, would be ready for Petunia. We were going to pull it off, as long as everything went as planned.

Chapter Eight
LILAH,
AN INTERMISSION

On Thursday, July 7, 2005, I decided to do something that I almost never did—stay home to get work done without the distractions of the people who kept stopping by my office to check on plans for Santa or Easterbunny or Xena, or to chat about nothing in particular. I had about one more day of work to go before I was finished with the first scientific paper about Santa. According to my calculations, Petunia was due within the next few weeks, so I wanted to get the paper out in the next day or two, just to be safe.

Rumors were already starting to circulate within the astronomical community that we were onto something big, and publishing the announcement about Santa quietly seemed like a safe way to deflect attention from the *real* big announcement that was soon to come.

The Thursday that I stayed home, Diane was at work in what was to be Petunia's room, putting some last touches on the decorations and furnishings, but I noticed little, since I was deep into the analysis and the explanations in my head. Still, at some point I noticed an unusual groan/sigh from the other room.

"What was that?" I called out to Diane.

"I'm just having a little cramping today. The doctor said I was supposed to expect something like this," said the ever-cool Diane.

"Are you sure? I've never heard you make sounds like that before."

"Nah. Just what the doctor warned."

I suggested that, for fun, we do a labor dress rehearsal. I would write down when Diane had little cramps and time them just as we would the real thing.

"Fine," Diane said, humoring my usual need to assign concrete numbers to everything going on around me.

I went back to work, a little more distracted now.

Fourteen minutes later, I heard the sound again. I remembered my birthing classes. Fourteen minutes was a pretty long interval. There was nothing to worry about. I was not even supposed to really begin timing things until the contractions were less than ten minutes apart. Even then, if contractions are more than five minutes apart you probably have many hours to go. The best thing to do is go on with whatever you should be doing instead. Like finishing a paper.

Diane nonchalantly replied, "Well, you missed the one in between."

What? That would make the contractions six minutes apart. And six minutes later, there was another.

"Umm, Diane? Could this be for real?"

Diane didn't think so but suggested that we pack our bags just in case if it would make me happy.

By the fourth contraction, I decided that I needed to be keeping track down to the second and I also needed to count the length of each contraction. Five minutes and twenty seconds, lasting fifty-one seconds. I started writing down the strength: stronger, really mild, strong, supermild.

We spent a few hours trying to decide if Diane was in labor or if this was just a false alarm. I plotted some graphs. The contractions came a little closer together, just as they were supposed to. And then they didn't. I trusted the expert opinion. We weren't supposed to do anything until they came four minutes apart, lasted a full minute, and did that consistently for an hour. But they never hit that magical four-minute interval. And they only occasionally lasted a full minute. And sometimes they were hardly there at all. The last contraction that I recorded that morning occurred at 11:14:40, four minutes and fifty seconds after the previous one. It lasted fifty seconds and was medium strong. Things were really going nowhere. And then Diane's water broke, perhaps changing astronomical history.

We were as calm as parents-to-be in labor could be, even stopping at my favorite coffee spot on the way (it was on the way, really; our birthing teacher had even recommended doing so!), since I knew it would be a long night for me. In the end, I didn't need the coffee, since there would be no all-night exhausting labor. Petunia had fooled us all by deciding to try to emerge foot first, and when the doctors realized this, they quickly whisked her out at the point of a knife.

"It's a boy!" the doctor assisting our primary doctor said.

A boy? How could we have gotten *that* wrong? I suddenly started thinking through everything that would be different.

"Girl, I mean." The umbilical cord had been strategically placed to cause initial confusion.

The three of us slept in the little hospital room that night, and in the still, slightly dark early morning, as Diane got some rest, I took the little swaddled bundle with me out into the hallway and stood in front of a full-length window looking east. It was less than three weeks past summer solstice, and the sun was making an early rise into an orange sky.

"Welcome to the world," I told Petunia. "The sun rises just like this every morning. It'll do it tens of thousands of times for you."

She opened her eyes and made a peeping squawk. Time for food, and I was of no use.

Petunia needed a real name. Diane used to joke, to anyone who would listen, that I'd get no say in her naming.

"Do you think I want her named Quaoar?" she would say.

We had tried to agree on a name a few months earlier. We had each made a list. Diane had crossed off all of the names on my list, and I had crossed off all of the names on hers. By the day after Petunia's birth, we still had no name. The nurses wanted a name for the birth certificate. In a sneaky move, Diane pulled a name from her list. For some reason, it seemed new and fresh. It means "night" in Arabic. And I didn't know anyone with that name. So Petunia became Lilah. And I can't imagine my world without a Lilah in it.

Those early weeks were a blur. Like most new parents, I slept no more than two or three hours at the longest. How tired was I? One morning I piled a load of laundry into the washing machine, scooped a plastic cup of laundry detergent from the box, and poured it into the receptacle in the washing machine. The detergent filled the receptacle and then spilled over the edges. This had never happened before. I had never scooped out more detergent than could fit into the receptacle. I thought hard. I stared at the detergent. I stared at the object in my hand. It was not the small detergent scoop, but a big plastic cup. Why would there be a big plastic cup in the detergent box? I read the side of the detergent box, then it became clear that this was not detergent but kitty litter. I had just loaded the washing machine full of kitty litter. I pondered what would happen if I started the washing machine with the kitty litter inside—the clumping kind!—and then spent the next thirty minutes trying desperately to get every last bit of litter out of the machine. Then I went to get some sleep; I could do laundry later.

Lilah did little more than sleep and eat and cry, which to me was the most fascinating thing in the entire universe. Why did she cry? When did she sleep? What made her eat a lot one day and little the next? Was she changing with time? I did what any obsessed person would do in such a case: I recorded data, plotted it, and calculated statistical correlations. First I just wrote on scraps of paper and made charts on graph paper, but I very quickly became more sophisticated. I wrote computer software to make a beautifully colored plot showing times when Diane fed Lilah, in black; when I fed her, in blue (expressed mother's milk, if you must know); Lilah's fussy times, in angry red; her happy times, in green. I calculated patterns in sleeping times, eating times, crying times, length of sleep, amounts eaten.

Then, I did what any obsessed parent would do these days: I put it all on the Web. It's still there, at least until Lilah gets old enough to find it and is sufficiently mortified that she makes me take it down (www.lilahbrown.com). I wrote thoughts about Lilah's sleeping and eating progress daily. Lilah developed an international following of people who, for whatever reason, were fascinated to know when a random infant would sleep and eat and what her father would say about it. If I ever missed posting data for a day, I was sure to hear about it from Lilah's fans.

In the years since, I have gotten occasional comments from parents-to-be or recent parents who stumble on Lilah's page. My favorite was from a first-time father in England who said that he kept the plot of Lilah's eating and sleeping posted on his refrigerator at home for the first six months of his daughter's life. He said it was vital to his sanity in those first few sleepless weeks to look up and see that, indeed, someday his daughter really would sleep more than two hours a time at night. My second favorite was from a friend who informed me after the birth of his daughter that reading Lilah's page was the worst thing he had ever done. His daughter was so much better than mine in every way that he had terrified himself for no reason. He had no statistical evidence to back up his assertion, so I, of course, didn't believe him one bit.

Today, when I go back and look at my comments and at the eating and sleeping records that I posted from July until the following March, when I finally ran out of steam, I can almost pull myself back into the moment. But mostly, I look at those early months and think: Really? Diane and I really woke up to feed Lilah three or four times a night for weeks on end? And feeding Lilah really took forty-five minutes? Twelve times per day? How did we have time for anything else? I think the answer is: We didn't.

With all of the plotting and charting, I thought I could do a better job of understanding things. And, of course, with understanding comes control. And if there was anything that I wanted to have some understanding and perhaps control over, it was Lilah's sleeping. Looking through Lilah's first six months, though, is a reminder that my understanding was minimal and my control was nil. But it didn't keep me from trying, plotting, calculating, predicting, and being continually proven wrong. Nothing could have been better.

Case in point, from my posting on the thirty-fourth and thirty-fifth days of Lilah's life, revealing almost all of my obsessions in only two paragraphs (which was all I could muster over two days, owing to sleep deprivation and lack of brain functionality):

Day 34 (10 Aug 2005): Lilah is my hero! She had her first more-than-5-hours-between-feeds day ever last night. We fed her at 9:50 PM and she didn't wake up until 3:10 AM, when I got up to give her a bottle, at which point she stayed asleep until 6:15. OK, so here's the math: Diane could have slept from about 10:30 PM, when she finished feeding Lilah, until about 6:05 AM when she started getting up to feed Lilah. That is more than $7\frac{1}{2}$ hours in a row! The reality is not quite as good, of course, as Diane seems to wake up whenever I get up to feed Lilah and when I come back and when the cats jump on the bed and when Lilah makes funny sounds. But let's forget these sordid details momentarily. This is probably the longest Diane stayed in bed at night in *months,* given that since about June, Lilah was pressing on her bladder and making her get up every several hours all night long. Fabulous! Way to go Lilah.

Day 35 (11 Aug 2005): I guess the job of heroes is to disappoint their admirers. Pretty mediocre night last night, including a pretty short sleep after I gave her a bottle, which leads me to question: does she sleep better when Diane feeds her than when I give her a bottle? I have always thought so, but the data do not support my suspicion. If you examine all non-bottle feeds between 1 am and 4 am the average interval between feeds is 2 hours 39 minutes. If you examine the same bottle feeds you find an average of 2 hours 28 minutes between feeds. Hmmm. Eleven whole minutes. And a Student's T-Test shows that this difference is not even statistically sig-

nificant. So I guess I should be happy about that. Me and a
bottle are almost as good as the real thing.

Besides Lilah's sleeping, my other main obsession was her
eating. Early on, I had taken over one of the nighttime feeding shifts, in an attempt to allow Diane to reclaim a small de-
gree of normality. I used milk that Diane had diligently—if
uncomfortably—expressed. I was in charge of milk stockpiling.
Diane would hand me a bottle with a bit of milk in the bottom.
If I thought we were going to need it soon, I would leave it fresh
in the refrigerator, but in times of plenty, I could invest a bit in
the future and put it in the freezer. Milksicles, we called them.

Over the first two months, as the milksicle bank began to
grow, Lilah and I ventured farther and farther away from Diane.
We went on hikes where I carried a tiny cooler full of ice packs
and frozen milk, and I would calculate just when I needed to
take out a container to thaw it so it would be ready precisely
when Lilah would be hungry. I would pay dearly for any mistakes. Milk not ready yet when Lilah was hungry? Lilah:
Waaaaaaaaaa. Brought too little, and she was still hungry? Lilah:
Waaaaaaaa. Brought too much, and some thawed and went to
waste? Me: Waaaaaaaa.

I invented—in my head—myriads of new devices specifically
designed to help parents acquire, manage, and efficiently use
their frozen milk supplies. One day I even started to create a sup-
ply database to record the comings and goings from the refrigerator and the freezer. Diane made me stop. "You're nuts," she
said. "Don't you have better things to do?" I did. I really did. But
I kept charting, graphing, and posting, nonetheless.

My last post was on March 4, 2006—day 240 of Lilah's life.
By March I finally was back to work full-time and Lilah was
spending her days with a nanny and another girl exactly her age,

who to this day remains her inseparable best friend. I was just coming back from a trip to the East Coast, where I had spoken about planets, new and old. But all I could think of at the time was what Lilah might be doing:

I've missed Lilah for the past few days. I'm on my way home from one of the longest trips since her birth. What's she going to be like when I get home? Actually standing? Able to wave bye-bye (we're working on that one now)? Finally relaxing now that mom and dad are more relaxed? Is that first bad case of diaper rash all resolved (we don't really need to talk about that, now, do we?). Can't wait. Can't wait. Can't wait.

And then that's all. I'm sure I didn't intend to stop forever that day. I'm sure I just got busy and skipped one day. Then two. Then a week. And then it was over. I'm sad now because as the memories have faded I can no longer go back and relive all of those moments of that time of Lilah's life. If I could, I would. I would do it all over.

Chapter Nine
THE TENTH PLANET

On the morning of the twentieth day of Lilah's life, only a few days after dumping kitty litter into the washing machine, I received a strange e-mail. A NASA official in Washington, D.C., wanted to know about Santa, which he called K40506A, the name my computer program had automatically assigned it on the day of discovery (*K* for Kuiper belt, 40506 for 2004, May 6, and *A* for the first one found that day). A colleague across the country was interested in studying K40506A, and the NASA official wanted to know when we were going to publicly announce the discovery.

My sleepy brain tried to make the connection: How would someone at NASA know about Santa, and, stranger, how did he know to call it K40506A? Had I told someone about it in the past few weeks? I couldn't remember mentioning it to anyone. Baffled, I did a quick search through my e-mails since Lilah's birth. Nothing but back-and-forth baby news and pictures. But the e-mail did jar my brain enough to remember that sometime in late July (and wasn't it now late July?) an online announcement would be made of the titles and subjects of hundreds of talks that would be given at an international planetary science conference in September. And near the middle of that list of talks were one by David and one by Chad, each talking about something that they called K40506A and which they declared to be the brightest object in the Kuiper belt. I, being on family leave, had no intention of attending any conferences anytime soon, but I was nonetheless listed as a coauthor on both of their talks.

I checked online and sure enough, the titles had been posted a day or two earlier, and people were already poring over them to see what we—and everyone else—were up to in advance of the actual meeting.

In the late afternoon, I wrote back to the NASA official and the distant colleague and said that we planned an official announcement of K40506A at the meeting in September, but that if it would be helpful for their research (and they could keep a secret), I would be happy to share the coordinates of the object earlier. I tried to be smooth and wrote:

> We weren't planning on making much of a big deal about this one. The mass is 32% that of Pluto, based on an orbital solution of its satellite. But we figure people are tired of hearing "almost as big as Pluto." We're waiting now for "bigger than Pluto."

Waiting, indeed. We were still months away from the planned announcement of Xena and Easterbunny, but the wait would be much shorter than I imagined.

I spent the next forty-five minutes cooking dinner, washing dishes, and putting real laundry detergent in the washing machine. Lilah woke up from her nap. Diane fed her. Lilah went back to sleep. I fed Diane. Diane went back to sleep. I fed myself. I was about to go back to sleep but instead checked my e-mail again.

An even stranger e-mail this time, from a colleague with whom we had already shared information about Santa so that he could help us with some of our ongoing studies. All he wrote was:

Mike, is this one of yours?

What then followed was a list of dates and positions in the sky of the location of an object discovered a day or two earlier by—by whom?—a name I didn't recognize, at a telescope I had never heard of.

My brain clicked in a little as I scanned the sky coordinates on the list. I'm not the type of guy to memorize coordinates of everything in the sky, but I knew that Santa was high in the midnight sky in about the April time frame. So were the coordinates on the list. I knew how bright Santa was; the brightness of the object on the list agreed closely.

My sluggish brain was now trying to accelerate to full speed for the first time in nineteen days. I did some quick calculations to get the precise coordinates of Santa on the days on the list, and I compared. Perfect fit. Santa had been found.

I remember this moment as a sharp pain in my stomach. We had been scooped. After discovering Santa six months earlier, and working hard to do a thorough job and write a scientific paper on the discovery (and failing, by one day, because of Lilah being born just a bit earlier than I had expected), someone had come out of nowhere and kicked us in the gut.

Who were these people, and what right did they have to take

my objects? My objects! Santa had been my baby for six months already. I looked up the culprits. I had never heard of them. They were at a small Spanish university, and they had never discovered anything previously. How could this have happened?

Chad sent an e-mail; someone had told him, too. He wrote:

Someone found Santa and beat us to the discovery!

There must be a way to make this all go away, I remember thinking. Maybe we could explain that we knew about it first. Or that our talk title was our announcement, our proof that we had been there first. Maybe there was still some way to salvage our discovery. There had to be a way. I was exhausted, but I knew that, with some sleep, I could find a way.

I heard Lilah cry. Diane was still trying to take a little post-dinner nap, so I let her sleep and went in to check on Lilah. I put on some music and danced with her for a while in her room. Just a few days earlier she had started making real smiles. She made one then. I sat down in the rocking chair with her until we both fell asleep. A few minutes later I opened up my eyes, put her down in her crib, and went back to my chair.

I had figured out a way to make it all right.

I sent e-mails to Chad and David telling them the details. I sent an e-mail to the NASA official with whom I had promised to confidentially share the position of Santa, saying that there was no longer any need for secrecy. And I started answering inquiries from the press who had seen the announcement and were already starting to take notice. They wanted comments from the guy who usually found these large objects out in the Kuiper belt, and they wanted to know how someone had beaten me to it.

Diane woke up and came in the room, and I told her what had happened. She protested that Santa had been my discovery, and I explained to her that no one owns the sky. If someone points a telescope at something, sees it, and announces it for the first time, it is that person's discovery, even if I knew about it earlier. In science, the first to announce takes the prize. The Spanish astronomers had announced Santa first, therefore they were the discoverers. Not only was there no argument that we could use to say otherwise, I didn't *want* to argue otherwise. I think the system is a pretty good one, even when it means I get scooped.

Perhaps this was even a good thing, I explained to Diane. In a few months we would be announcing Xena and Easterbunny—both even bigger than Santa—and having an earlier announcement of a large object from a different group in a different country on a different continent added a bit to the excitement of it all. I could not have come up with a better plan myself.

Diane, without the benefit of the adrenaline that had been pumping through my system for the previous hour, stared at me as if I were a lunatic. But as lunatics go, I was not too crazy. With the question of who discovered Santa now settled, we might as well make something good out of it.

Diane went back to sleep, while I went back to my e-mail.

The astronomical-media grapevine had picked up on the fact that K40506A—the object that Chad and David had included in the titles of their talks at the conference in September—was the same as the new object just announced (which now had yet another name: 2003 EL61, based on the fact that the astronomers who had discovered it found it by looking through old images from 2003, much as I had been looking through old images myself when I found it). A headline from the BBC blared: "Conflicting Claims in Planetary Discovery." The article breathlessly exclaimed how astronomers were already passionately arguing about whose claim was legitimate and how the dispute was bound to reach the highest levels of the International Astronomical Union. And that the object could well be twice the size of Pluto.

Twice the size of Pluto? We knew, of course, that 2003 EL61 or Santa or K40506A or, later, Haumea was only about a third the mass of Pluto. We had followed the orbit of Rudolph, the tiny moon going around Santa, and had accurately determined the mass in the process. But the new discoverers didn't know anything about the moon. They had discovered Santa/2003 EL61 only a few days earlier and hadn't taken the time to do anything but make the announcement. Nobody knew about the little moon, because I had never quite finished the paper announcing its discovery.

I suddenly acquired a new worry. If the press started gushing about something potentially bigger than Pluto that turned out to be only a third the size of Pluto, what would happen in a few months when we announced the existence of something that really *was* bigger than Pluto? Would people simply say, "Oh yeah, we heard about that one already"?

With the perspective of the years that followed, and with the help of a reasonable amount of sleep, it is now clear to me that my worry was misplaced. Things that are real, that are important, will go into textbooks, into documentaries; they will become part of our culture. Everything else will fade. Still, at the time, I thought it very important to do two things. First, I wanted to make sure that no one could possibly claim that I was attempting to steal any credit for the Spanish team's discovery, and second, I needed to make sure that everyone knew as quickly as possible that 2003 EL61/Santa was only about a third the size of Pluto.

First, to quickly answer all of the e-mail questions I was getting from reporters, I made a website describing 2003 EL61 and what we knew about it. I put a picture of Santa and its moon Rudolph on the site, showed the orbit, and explained how we knew that Santa was only a third the mass of Pluto. I described how we had found it back during the previous December and were preparing a paper describing the discovery. And then I wrote extensively about why I thought that the accepted scientific practice of assigning discovery rights to the first person to announce was the right thing to do.

Why was it the right thing to do? It is the only way I can think of to strike the right balance between the desire of the broad community to have all information be public immediately and the desire of the individual to keep a discovery secret for years while slowly studying all of the implications and making all of the important findings before anyone else gets a crack at it. Both of these are natural desires. Neither of these is a particularly good idea. Instant disclosure leads to unvetted science being thrust into the community (such as the claim that 2003 EL61 was twice the size of Pluto), and it leads to a lowering of the incentives to make the discoveries in the first place. On the other hand, keeping discoveries secret prevents the broader scientific community from learning even more about the discoveries.

It took science a while to settle on the present system. In 1610, while watching Venus with his new telescope, Galileo noticed that it went through phases identical to those of the moon. He knew that his discovery was big, and he wanted to make sure everyone knew he had found it first, but he also knew that the discovery would be even better if he could wait just a few more months for Venus to come around to the other side of the sun and show the opposite phases. I understand how he must have felt, waiting for his planet to emerge from behind the sun; Xena still had another month to go before I could finally see it with the Keck telescope. To prove that he had already discovered the phases of Venus, Galileo wrote to Kepler, "Haec immatura a me iam frustra leguntur oy," which translates as something like "This was already tried by me in vain too early." In case anyone else later claimed to have been the first to discover the phases of Venus, Galileo would be able to point out that his note to Kepler was really an anagram for *"Cynthiae figuras aemulatur mater amorum,"* or "The mother of love imitates the shape of Cynthia": The mother of love—Venus—imitates the shape of Cynthia, the moon. The fact that Venus goes through phases just like the moon instantly proves that Venus goes around the sun, not the earth. Two millennia of understanding of the universe around us had to be thrown out the door at that moment.

These days anagrams don't count. You have not officially discovered something until the moment of a scientific announcement.

I explained all of this on the website I put up overnight. Even so, I was tempted to add an anagram of my own: "The neat white elephant enthralls," which rearranges as: "The tenth planet is near the whale," which obviously refers to the fact that Xena is in the constellation Cetus, the whale. Unlike Galileo, though, I resisted. I was going to take my chances with Xena and Easterbunny.

Sometime in the late evening, I noticed an e-mail from Brian Marsden—gatekeeper of the solar system—with whom I had interacted on all of the other discoveries. He found the Spanish discovery suspicious, coming the same day as the name K40506A had appeared in public. He wanted to know if there was any way that the Spanish could possibly have been able to figure out where Santa was simply from knowing the name K40506A.

No chance, I said. It would be as if I'd decided to nickname some city somewhere in the world Happytown, and simply hearing the nickname, someone had picked up a globe and pointed to the right spot. No chance whatsoever, I told Brian.

Around midnight I started working on the second part of my plan. I wanted to make sure that no one thought I was going to stake a claim. I wrote directly to the discoverer, an astronomer named José-Luis Ortiz.

Dear Dr. Ortiz—

Congratulations on your discovery! We found the object, too, about six months ago and have been studying it in detail for the past few months. It has a few interesting properties that you might find interesting. Most interestingly, it has a satellite, and the orbital solution gives a system mass of about 28% of that of the Pluto-Charon system. It's still probably the biggest KBO around but it has a sufficiently high albedo that it is not quite as big or massive as Pluto. I've got a paper describing the satellite that, ironically, I was planning to submit tomorrow. I will forward the paper to you as I submit it.

I am sure that I will get inquiries about your new object from different people; is there [or is there going to be] a website describing your survey or your discovery that I can point people to?

Again, congratulations on a very nice discovery!

Mike

A critical analysis of my e-mail suggests several things. First, I repeat the word *interesting* a lot when I am tired. Second, I was amazingly generous for someone who hours earlier had been trying to figure out how to turn back time and claim the discovery. If Ortiz had heard the stories that we were going to put up a fight, he must have been quite relieved to get this friendly e-mail congratulating him on his discovery. Third, I parsed my words very carefully. We "found" the object, but Ortiz had "discovered" it, and I repeatedly called it "your" object. But I was not 100 percent straightforward. The claim that we were planning to submit the paper the following day was now true, but it hadn't been true until we'd learned of this discovery. Overall, when I look back on this e-mail years after the fact, I am proud of myself for having written it.

Lilah was awake for a late-night feed, and it was my turn. She drank quickly and went back to sleep, not aware that anything in particular was going on.

I had one more task that night before going to sleep. I needed to finish the paper about Santa. I dug out my notes from twenty days earlier and tried to remember what else I had left to do. Very little, it turned out. In only a few more hours, I had tidied up the manuscript, finished one more calculation, uploaded everything to the website of the scientific journal, and pressed "submit." The paper that was supposed to accompany the announcement of the existence of Santa was on its way, though it was now a paper describing an object called 2003 EL61 that had been discovered by someone else. I linked the paper to my own website so it could be found by all and crawled to bed as Diane was getting up to feed Lilah.

I didn't sleep. Brian Marsden's question about a link between "K40506A" appearing publicly and the Spanish discovery kept circling in my brain. Still, I couldn't imagine how the name K40506A could be used to discover Santa. Finally, I got up, went back to my computer, and googled "K40506A." First up were the stories now appearing overnight about the discovery. Second were the titles of Chad's and David's talks. Third was something strange: a list of objects that had been observed by a telescope in Chile one particular night in May, including an object called K40506A. And it told where the object was.

What was this?

The address of the webpage was long. I went up successive levels of the tree and finally realized that the list was a record from a telescope in Chile that David had been using to watch Santa; the list was not even from the telescope itself, but rather from an astronomer at the University of Ohio who had built and kept track of the camera that the telescope used. And one of the seemingly innocuous things that he'd kept track of was what the camera he built was looking at and when.

I looked further and realized that if I fiddled with the Web address I could change the table to display different nights. The object K40506A appeared again, but now with a new position. A slight panic rose in my stomach. I kept fiddling with the Web address and kept getting the object's coordinate on different nights. K40506A kept moving. Knowing where the telescope was pointing on successive nights as it tracked K40506A was as good as knowing where K40506A was on successive nights. And with successive nights of knowing where it was, it was only a small leap to knowing everything.

I didn't think anyone would have gone to such lengths to steal the positions of Santa, but I suddenly had a new worry. On one of those nights when David's Chilean telescope had watched K40506A, it had also watched K50331A and K31021C. I recognized those codes, too. They were Easterbunny and Xena. This was bad. Because the name K40506A was publicly involved in an astronomical controversy, people would *certainly* google the name, just as I had, and they would see a May position of K40506A. Some would take it a step further and fiddle with the

Web address as I had and find even more positions of K40506A. Some would even notice that other similarly named objects— K50331A and K31021C—occasionally appeared on the lists and wonder what they were. Some would track them down. And some would be aware enough about what it all meant to calculate positions in the sky. They would suddenly know exactly what and where Xena and Easterbunny were.

In the middle of the night, I sent an e-mail to Chad and David warning them about all of this and asking if they knew how to get the information off the website in Ohio. I then wrote to Brian Marsden and told him the news, too. You *could* use the name K40506A and a little Google sleuthing to figure out where it was. I then explained to Brian that while I was not paranoid enough to think that Ortiz had done this to find Santa—it was inconceivable that any astronomer would actually be that underhanded—I was *definitely* paranoid enough to think that now that Santa was out of the bag, someone would eventually find our two other objects the same way. I then told him about Xena and Easterbunny. I told him that our goal was still to wait a few months until we had scientific papers prepared on these discoveries before announcing them.

Finally I went to sleep. I slept through two successive Lilah feedings for the first time in twenty days.

When I woke up, I told Diane everything that had happened that night. I had coffee. I bounced Lilah around the house a little bit. Then I checked my e-mail again.

The press was fascinated, both by the larger-than-Pluto part of the story and by the astronomer-fisticuffs part of the story, even though neither was true. I kept pointing people to the webpage.

Ortiz wrote back, a little overwhelmed, it seemed. He

pointed me to a bare-bones website he had thrown together to describe the discovery. I added a link to his webpage from mine so people could read about the initial discovery.

Brian Marsden wrote back suggesting that perhaps I should be more suspicious. Did I not find the circumstances surrounding Ortiz's discovery and announcement of 2003 EL61 odd?

David wrote back and said there was nothing we could do about the webpage in Ohio.

At 9:18 a.m. I got a new e-mail from Brian. It contained a list of all of the telescope positions of Easterbunny and Xena. Someone had already found all of the positions on the website in Ohio and had sent them in to the place where you announce discoveries. At the same time that the coordinates were sent to Brian, they were also sent to the Internet chat group that had been angry with me about my naming of Sedna and Quaoar. All of the information was now public, and there was no possible way to contain it. We were going to have to make the announcement right then.

I wrote to Brian with the official data and told him to proceed with the announcement. I wrote to Chad and David and told them what had happened and that we were instantly going live. And I sent one more e-mail to Ortiz:

Jose—

Along with 2003 EL61, which you discovered this week, we have also been tracking two larger Kuiper belt objects. After the 2003 EL61 announcement someone tapped into an online database to see where we had been pointing our telescopes and, in doing so, reconstructed the positions of these two other Kuiper belt objects. They have now made these positions public. Because of all of this, we have had to announce these discoveries this morning.

I am very sorry that this announcement has to come the day after the announcement of your own discovery and that this will likely overshadow your very nice work. I will continue to try to make sure that you get the credit you deserve for the 2003 EL61 discovery.

Mike

Next, I needed to quickly make public webpages about Xena and Easterbunny, which would soon be getting new names. I would probably need to work on a press release. I kissed Diane and a sleepy Lilah goodbye and drove down to Caltech for the first time in twenty days.

At work, I called the Caltech press office and told the person who writes press releases, "We've discovered something bigger than Pluto and need to have a press release about it go out today."

"Bigger than Pluto!" he exclaimed. "Wow! So it's the tenth planet?"

I hadn't figured that part out yet. I had strong opinions about planets. I didn't believe that Pluto should be classified as a planet. The word *planet* should be reserved for the small number of truly important things in the solar system. Xena, though bigger than Pluto, did not rise to the level of a truly important object in the overall context of the solar system.

But but but but still.

"I don't want the press release to say it's a planet. Just say it's something larger than Pluto," I replied.

"Are you crazy?" he said. "This is the biggest astronomical discovery in the solar system in a century, and you're going to be the one arguing that it's not a planet?"

Uh. Yeah.

"If you call it the tenth planet, the public will be excited and engaged. If you call it the biggest not-a-planet, people will just be confused."

I remembered Diane's words from before Lilah was born. I gave in. The press release that went out to the world that day was titled "Caltech Astronomer and Team Discover the 10th Planet."

I would have a lot of explaining to do, once I got some sleep.

Next up was to arrange a press conference. I called my NASA contacts and told them that I needed to arrange a press conference that afternoon to announce the discovery of the tenth planet.

Impossible, they said. The space shuttle was up at the space station with missing tiles, and people were worried about a crack-up on the way down. They were having a press conference about *that* this afternoon. How about Monday?

Impossible, I said. If the announcement was not made before the sun went down, almost anyone with a modest-sized telescope could point to the now publicly available positions and say they had discovered the thing.

"Did you just say tenth planet?" they asked. We set up an international press conference for 4:00 p.m.

As soon as I hung up the phone, it rang. It was an old friend from college, Ken Chang, who happened to also be a science reporter for *The New York Times*.

"Tell me about this big object," Ken said.

"Which one?" I asked.

"Um, what?" he said.

He was calling about Santa/2003 EL61, of course. He had not yet received the press release about the tenth planet. I quickly told him about our big discovery and asked him if he could wait until the 4:00 p.m. press conference to get the details.

"Four p.m.? On a Friday afternoon? To announce the discovery of the tenth planet? Are you nuts?"

It seemed to be the question everyone was asking me that day. I hadn't even realized that it was a Friday, but that was good information to try to store in my brain. And, oh yeah, that it was July.

"Friday at four p.m. on the West Coast is too late for me," Ken said. "It'll miss the Saturday and Sunday papers and be old news by Monday."

I told him about the discovery. When he asked me the name of the new planet, I realized that I didn't even know yet what the official license plate number designation was going to be (it turned out to be 2003 UB313). I told him that it had no name yet.

"Well, what do you guys call it among yourselves?" Ken wanted to know.

"Xena. It will have a real name soon, but for now we call it Xena."

Ken chuckled and wrote it down.

Contrary to what I thought that morning, it would not get a real name soon. After Ken wrote it down that first time, Xena became its nickname for more than a year. There are many people, I believe, who still think that the object remains named Xena.

Ken Chang was right. The story did end up missing almost all of the Saturday and Sunday papers, and though the discovery was not exactly old news by Monday, it was indeed clear that Friday at 4:00 p.m. is not the right time to make a press announcement—unless, perhaps, you are announcing that you are going back to rehab, and you hope no one notices. But at least, by virtue of that one accidental phone call, the announcement of the discovery of the tenth planet hit the front page of *The New York Times* on Saturday, July 30, 2005.

By about noon on Friday, I had built a webpage describing Xena. It was spare but would have to do. I drove up to the Jet Propulsion Laboratory—JPL—where they had the facilities to put on a major press conference.

I can no longer put together a timeline for the rest of that day; most of the memories are simply too jumbled. I recall at some point changing shirts and shaving in the men's room at the press building at JPL. I don't remember a single thing I or anyone else said at the press conference, though I vaguely remember standing in front of a TV camera with a small speaker in my ear; every three minutes I was connected by satellite to some different TV show. I don't know what I said, and I certainly don't want to know how I looked.

I drove home late in the evening. A few minutes after arriving home, the head of the media department at JPL called me to double-check that I was all right. I remember that conversation extremely well. "I'm fine," I said. "I'm lying on the bed and Lilah is asleep in my arms. What could be better?"

"Good," she said. "Then would you mind doing *Good Morning America* on Monday morning, and they want you to bring Lilah."

At 2:00 a.m. on Monday, Diane, Lilah, and I drove down to a Hollywood studio. Normally I would consider this hour to be thoroughly indecent, but given the round-the-clock schedule we were currently on, 2:00 a.m. was no better or worse a time than 2:00 p.m. Actually, it was better, as there was no traffic.

When I arrived at the studio, I was hooked up with earpieces again and talked about planets, old and new, with Charlie Gibson and Diane Sawyer. At the end, my wife brought Lilah over for the cameras. Two thousand miles away, in Alabama, my mother was on the edge of her seat. She already knew about all of the planet parts, so that was just filler. But it was the first time she had ever seen Lilah.

According to my calendar, the following weeks were a storm of interviews and talks and TV appearances, of which I have no memory. If you look at the records I kept of Lilah's eating and sleeping and crying and smiling, you would not know that any of it had happened.

A week after the biggest scientific announcement of my life, it seems that all I cared about was whether or not Lilah would sleep and how frequently she would feed.

Day 31 (7 Aug 2005): Lilah is one month today! To celebrate her birthday she had a record sleep last night, almost 5 hours! It included an hour-long car ride at the beginning, which may or may not have contributed, but to top it off she then had two 3½-hour sleep sessions in a row. If you look carefully you will also note that she is, in general, stretching things out more (well, at least at night). For the past 5 days we have dropped from 10 feeds a day to 9 feeds a day. This may not seem like much to you, but it is about 45 minutes of saved time for Diane every day (or, more accurately, 45 minutes of extra sleep at night)! There was even her first 8 feed day back on Day 29 that originally passed without note. My original complaint back around Day 12, when it appeared that Lilah couldn't distinguish between day and night, is clearly no longer valid. Night times are definitely for longer sleep periods. Thank you Lilah, thank you thank you!

Chapter Ten

STEALING THE SHOW

The Internet chat group that had been irritated with me over the discovery and naming of Quaoar and Sedna was up in arms again. I didn't know it, but Ortiz himself was apparently an occasional member of this group, and many were rallying around to defend him against the onslaught of the evil American astronomers trying to deny him credit for his discovery. Except, of course, there was no onslaught. I told anyone who would listen that Ortiz had indeed discovered 2003 EL61/Santa. Since I couldn't really be excoriated for trying to steal Ortiz's credit, they would find something else to rail against. They then argued that I had made up the story that someone had found the coordinates of Xena and Easterbunny, so that I would have an excuse to hold a press conference the day after Ortiz's discovery in order to overshadow him. And then they hit on a new accusation: I was bad because I had been trying to keep Santa and Xena and Easterbunny secret. I chuckled and shook my head, given how hard we had tried to do everything correctly by scientific standards.

Even Ortiz got into the act, declaring in an interview:

> With technology many times more advanced than ours, Brown's team had discovered three big objects many months ago, but they were hiding its [*sic*] existence from the international scientific community, as they did before with Quaoar and Sedna.
>
> This secrecy was useful to Brown, as it allowed him to study his own findings in detail and exclusively. But his actions harm science and don't follow the established procedures, that imply notifying the existence of a new object to the astronomical community as soon as it's discovered.

Sigh. I almost sat down and wrote a long article on why the instant announcement of discoveries is precisely what good scientists *don't* do and that the established scientific procedures are to confirm findings and write scientific papers before making public announcements, but I decided that the accusations were sufficiently ridiculous that I should ignore them and let them fade into their deserved oblivion.

I will admit, though, to being stung and irritated to read Ortiz's comments. I didn't care about what nonastronomers were saying on chat groups, but I thought it harmful for professional astronomers to spout such nonscientific nonsense. And it seemed particularly uncharitable given how hard I had been defending Ortiz against all accusations and deflecting credit to him whenever possible. Odd, I thought.

Given all of the chatter, I decided to write to Ortiz again to assure him that I considered him the legitimate discoverer of 2003 EL61. I asked him if he had thought about what name he would like to give it. Only the discoverer is allowed to propose a name, so this was a pretty unambiguous signal of my intent. I told him that we would be interested in giving the moon that we had discovered a name that fit with the name that they proposed for 2003 EL61. Ortiz wrote back thanking me for asking but saying that because of the recent onslaught they had had no time to even begin considering a name.

The chat group continued to try to prove my malicious nature. One of the main proponents of this theme was the German amateur astronomer who had, a year and a half earlier, tried to thwart our naming of Sedna by naming some of his own objects Sedna. He had, interestingly, even taken part a bit in the Ortiz discovery. After Ortiz found the object in his old data, he had contacted the German amateur to get a current picture of the object. The amateur had promptly complied, becoming in the process a secondary member of the discovery team. It was an odd coincidence that the one person who appeared to have the biggest ax to grind against me happened to be involved in all of this. But coincidences happen all the time. I thought nothing of it. Brian Marsden, when he first learned this, said, "I smell a rat in here somewhere." Marsden, as I continued to learn, has an acute sense of smell.

Interestingly, after some time, a countertheme began to develop among the members of the chat group. Not everyone appeared to be convinced that the discovery of 2003 EL61/Santa had been legitimate on the part of the Spanish group, and they started asking Ortiz probing questions. One particular question interested me: Did Ortiz know about our discovery of Santa before he claimed that he had discovered it himself? Had he ever accessed the website with all of the coordinates? Ortiz never responded, though his friend the German amateur defended him viciously through counterattack and accusation. It was all quite ugly, though perhaps no more so than many other chat groups on the Internet these days. I figured it was best to stay out of the fray.

A week and a half after the initial announcements, I got a phone call out of the blue from an astronomer I didn't know. Rick Pogge was a professor at Ohio State University, and his website database was the one that had been tapped into, forcing us to make the sudden announcement of Xena and Easterbunny. He was apologetic about what had happened. I told him not to worry; it would not have occurred to us or to someone else that anyone could have figured out a way to use these generally dull databases for nefarious purposes. And it would have been even less likely to occur to us that someone would actually do it. He then described all of his recent changes to the database, explaining how this sort of thing would never occur again.

"Great," I said. "That sounds great."

"But there's something more that you need to know," Rick said.

More?

Rick then told me an interesting story.

He, like everyone else, had first learned about Xena and Easterbunny when he'd read the accounts of the press conference a week and a half earlier. As scattered press reports came out about someone tapping into some database, Rick first thought it was a really unfortunate story; then he thought, Wait, is that *my* database? Indeed, Rick had built the camera that was mounted on the telescope in Chile that we had been using to monitor Santa, Xena, and Easterbunny. One particularly nice feature of that telescope in Chile was that for routine observations, like taking pictures of the positions of our Kuiper belt object, we didn't have to fly to Chile each time we wanted a picture, but instead a person permanently stationed in Chile would take the pictures we needed using the camera that Rick had built. Rick then maintained the database of observations that allowed astronomers to access their pictures after the camera had taken them.

After suspecting that perhaps it was this database that had been tapped into, Rick became curious and began to look through the computer logs to see who had accessed the database. In the years that the database had been up, it was accessed almost exclusively by people who were supposed to be accessing it: the astronomers who were using the telescope that the database related to. Occasionally inadvertent access would show up once and never again.

But the records also showed that one day in late July something odd had happened. A computer address that Rick didn't recognize accessed the database multiple times in quick succession. Each time it accessed the database, it was pointing to a different webpage that showed the location of an object named K40506A on different dates. Rick looked up the computer address to see where it was from. It was from Spain. He looked in more detail. It was from the institute in Spain where Ortiz was a professor. This access to the database occurred two days before Ortiz announced the discovery of 2003 EL61. Ortiz had known all along.

I sat at my end of the phone, stunned. I had Rick go back and tell me precise dates, times, and the computer addresses, and I wrote them all down.

There was more.

On the first day that Ortiz had tried to announce the discovery, he had inadvertently sent the announcement through the wrong channels, so he received no reply. The next day, he had sent a much more thorough announcement, including new observations by his German friend and more data from other old images. All of these extra data would have required knowing the position of the object more accurately than before. The morning before Ortiz sent all of the old data, Rick's database had been accessed once again. A quick flurry of websites had been viewed, each showing the position of K40506A on different nights.

I kept writing. I was going from feeling stunned to feeling slightly giddy. The Spanish guys *had* stolen Santa out of the database, but they had botched the job. There were fingerprints all over the scene of the crime. And now they were busted.

After I hung up the phone with Rick Pogge, I immediately called Brian Marsden.

"I knew it," he said.

All I knew from Rick was that the computers accessing the database were at Ortiz's institute in Spain. But Brian had an interesting idea. "Tell me those computer IP addresses," he said. He then cross-checked them with e-mail he had received. The specific computer that had accessed the database the first time was the same computer from which the initial announcement was sent. The specific computer that had accessed the database the second time was the same computer from which the second announcement was sent. The first e-mail had come from Pablo Santos-Sanz, a student of Ortiz's, while the second e-mail had come from Ortiz himself. The fingerprints matched perfectly.

Though I will likely never be able to confirm most of this, here is my hypothesis as to what actually happened:

On the second-to-last Wednesday in July, the titles for talks to be given at the big international conference were announced, including talks by Chad and David, which mentioned K40506A and described it as big and bright. The following Tuesday, Santos-Sanz noticed the titles, and, curious about K40506A, he typed it into Google. He was likely shocked (as I would be a week later when I did the same thing) to find precise information about where a telescope was pointed one night in May. After the initial shock, he must have felt some nervous excitement. He must have been savvy enough to realize that he might be able to find more information about where the telescope was pointed. He must have looked at the Web address and realized that it looked something like

www.astro.osu.edu/andicam/nightly_logs/2005/05/03

and he must have made the quick assumption that the last bit was the date. He changed it to something like

www.astro.osu.edu/andicam/nightly_logs/2005/05/05

and was suddenly rewarded with the position of K40506A on a different night. He collected a few more positions and set to work. Knowing precisely where the telescope was pointing over multiple nights is precisely the same as knowing where the object is on multiple nights. And knowing *that* means that you know enough to go find it yourself.

What happened next I cannot figure out. Here is the story as I envision it. I *think* that Ortiz and Santos-Sanz really were engaged in a legitimate search for objects in the Kuiper belt, even though they had not yet been successful. My guess is that they had never gotten around to writing the computer software to help them with their search, so they merely had a big pile of images dating back several years, with no way to look at them. It wouldn't be surprising. As I had learned over the past few years, writing the computer programs to analyze the data is at least as hard as collecting the data itself. But armed with the previous positions of K40506A, Santos-Sanz no longer had to look through all of his images; he could quickly determine which ones might have the object on it, and he no longer needed to write complicated software to look through a vast pile of images. He could instantly go to the right images—the ones where he knew K40506A had to be—and do a quick search by hand. He found it. He showed Ortiz. They announced their "discovery" on Wednesday, thirty-eight hours after the first data access. They must have had a busy thirty-eight hours.

When the initial announcement received no acknowledgment (having never discovered anything before, they were unclear on the proper methods of sending in a discovery), they must have decided they needed more images to demonstrate that it was real.

At this point, it remains possible that Ortiz was in the dark about what had happened. Perhaps Santos-Sanz had not told him about the computer access. Perhaps he was going to try to make it appear as if he had gotten all of his software written after all and had made a quick and spectacular find. But on Thursday morning, the day they decided they would need more images to convince people that their discovery was real, the database was accessed again. This time the access came from Ortiz's own computer. He did the same tricks to find more positions. Twelve hours later, Ortiz's German amateur astronomer friend—the one who passionately disliked me—was observing the object from a telescope in Majorca. Two hours later, Ortiz re-sent an announcement of the discovery including the images from that very evening, in addition to old archival images that the German amateur had tracked down for them.

This time the announcement went through the right channels. I would find out about it a few hours later, on a Thursday afternoon, while home with Diane and a twenty-day-old Lilah. Seven hours later, I sent my e-mail to Ortiz congratulating him on his fine discovery, thinking he had discovered something in the sky, not in the bowels of the Web.

Brian Marsden had two more questions for me: What about the German amateur? Surely he was involved in this somehow. I told him no. Only the Spanish computers had accessed the database. I was certain that if the German amateur had learned about the computer logs he would not have been able to resist looking at them himself. I suspected that he had been duped like the rest of them. And he had been duped so well that he felt it right to be a vicious defender of the honesty of Ortiz.

Brian's last question: What are you going to do about this?

I didn't know. I hung up the phone. My anger was beginning to grow. These guys had stolen our discovery and, what seemed even worse, forced us to make an incomplete and hasty announcement of the biggest astronomical discovery of my lifetime. They had caused me to spend most of my past week at work rather than at home, where I was supposed to be on family leave. And these guys would have gotten away with it, too, if not for the careful sleuthing of Rick Pogge. What would be the right response? Public humiliation? An interstellar smack-down? I decided that, for now, the main thing I needed to do was go home.

Diane and Lilah were home. The three of us sat in our favorite resting spot—Diane and I lying in opposite directions on the sofa, our feet intertwined, Lilah alternately resting on one or the other of us.

I told Diane what had happened that day.

"What are you going to do?" she asked.

I still didn't know. I was tired. I was angry. At one stage in my life I suspect that I would have gone immediately on the offensive and publicly blindsided Ortiz with what we knew. It would have been a thorough and exceedingly satisfying public crushing. It was certainly what I believed that Ortiz deserved.

But I didn't do it. At least not yet.

Why not?

Try as I may, I can't put myself back into my state of mind that day. I can't remember exactly all of the different things going on. One thing I can do, though, is go back to look at Lilah's website from the day.

After sitting on the sofa with Diane that evening, Lilah was being fussy and wanted more attention from her parents. I got up to put her in her crib. When she finally got to sleep, I sat down at my computer and wrote a post on Lilah's site.

It went like this:

Day 33 (9 Aug 2005): Approximately 7:30 pm. Lilah was crying immediately after being fed and I pulled out my best move, which is to dance with her to Jack Johnson, "Better When We're Together" around the living room. She cannot resist falling asleep to this. Except that halfway through the song I found myself inconsolably bawling and thinking that in another two or three or four decades I might be dancing with Lilah at her wedding to a song much like this. Hey, Lilah, by the way, if you are looking for a good father-daughter dance song at your wedding, my vote, as of your fifth week of life, is this song. I've got it on this thing that we use these days that we call a CD. I'll explain all the old technology to you some day. I hope it still plays OK over the dried-up tears.

And then I wrote to Ortiz:

Jose—

As you can well testify, I have been quite supportive of your announcement of the discovery of 2003 EL61, and I have tried to make it as clear as possible in all public pronouncements that I regard your discovery as 100% legitimate.

Given this support, I am now extremely disappointed to learn that you have been less than honest about your actions. We have examined the web logs to the SMARTS records and have found that your computers examined those records shortly before your announcement of the discovery.

I regard this as a serious breach of scientific ethics and will make this information public shortly, but I would like to allow you the possibility of responding first. If you would like to in any way explain your actions please let me know within the next day.

Sincerely,

Mike

I went to sleep.

The next day, I checked for a response: nothing. And nothing the next day. And the next. I had said that I was going to make all of this information public in the next day or two, and the time was up. What should I do?

I waited. I couldn't do the public blindsiding. I realized that I wasn't looking to crush or humiliate Ortiz. I just wanted him to admit what had happened and to say he was sorry.

I waited. But my patience was not infinite. I wanted my apology. And I needed Ortiz to understand that his breach was serious. After weeks of silence from Ortiz, I wrote to the director of the institute where Ortiz worked:

Dear Dr. del Toro—

I regret to have to inform you about a formal complaint I recently made to the IAU about what is apparently unethical behavior by Jose-Luis Ortiz. As you are no doubt aware, Dr. Ortiz reported the discovery of the bright trans-Neptunian object 2003 EL61 late last month. At the time many in the community questioned whether Ortiz had found out about the object first from perusing our observing logs for the same object. I have repeatedly said that I believed that Ortiz's discovery was legitimate and that I supported Ortiz et al. as receiving sole credit for the official discovery. I have publicly and privately dismissed the accusations and congratulated Ortiz and his team.

Sadly, I now find that Ortiz has not been honest about his access to our observing logs. We have now examined the records for the web server which show that Ortiz and Santos-Sanz indeed did access our observing logs. The first access came two days before the announcement of the discovery. Our observing logs were accessed multiple times over multiple days. Logs from several different nights were accessed, allowing a complete calculation of the orbits of all of the objects that we were tracking.

I feel it is likely that the explanation that will be proffered by Ortiz will be that the access to our observing records occurred after they had already made the discovery. . . . This is still a serious breach of scientific ethics. They accessed our observing logs, checked the object we were observing, noted it was the same as theirs, and quickly rushed to announcement with no acknowledgment of having known of our previous observations. I believe that such a behavior is a serious breach of scientific ethics and is deserving of censure.

It seems equally likely to suppose that Ortiz knew nothing of the object until accessing our logs. If this is the case, the behavior amounts to scientific fraud and is deserving of termination.

I have attempted to contact Ortiz for an explanation, but have received no response for almost 3 weeks now. I would have preferred that he have a chance to state his side of the story, but I am unwilling to wait any longer to make the record of his actions public. I intend to publicly post the detailed timeline of his access to our observing logs early next week.

I hope that you feel, as I do, that these allegations of potential fraud at your Institute are extremely serious and should be thoroughly investigated.

Sincerely,

Mike Brown

The director must have quickly realized the seriousness of the charges. He promised to gather information, and he begged that we not hold the actions of Ortiz against his institution as a whole.

In a second e-mail, the director informed us that he had spoken with Ortiz and encouraged him to respond.

Having now had most of the month to think about all of the implications of what had happened, I wrote a much more detailed e-mail to Ortiz and Santos-Sanz, with a full accounting of everything we knew. It was hard for me to imagine that there was any way out for them but to confess and apologize.

I waited.

After nearly a month, I opened my e-mail one day to find a message from Ortiz. What would his response be? Would he angrily deny his actions? Would he gush gratefully for having been given the chance to redeem himself after such a mistake? Would he try to negotiate some solution favorable to himself? I was eager—and nervous—to see what his response was going to be; it was likely to thoroughly define whatever course the future was going to take.

Ortiz came out swinging: I was to blame for all of this, and indeed, I was universally regarded as a menace to science itself, so I should be the last one to discuss ethics. Instead, I should apologize to the international scientific community and quit my secretive ways. If I reformed, Ortiz was even willing to give me credit for the discovery of 2003 EL61. He would be happy simply to be noted as the person who first reported the discovery. I should think it over, and we should talk again at the end of the month.

It seemed to me that in Ortiz's view he had not stolen K40506A; he had liberated it. I had been hiding it all along, in clear violation of what he considered accepted rules of science. Thus Ortiz should be commended rather than condemned for taking the information and setting it free.

In some ways that he didn't quite know, Ortiz was right. I *had* been hiding information. I had detailed records of exactly what Ortiz had done, and I had delayed telling anyone for a month, in the hope that an amicable solution could be found. That hope was now clearly dashed. I posted the records of Ortiz's access to our database on the Web. The next day, a long story about allegations and counterallegations appeared on the front page of the science section of *The New York Times*. Stories about the allegations showed up in all of the major scientific newsmagazines. And on the day that the story broke, José-Luis Ortiz was named Worst Person in the World by Keith Olbermann of MSNBC, beating out a Sri Lankan Airlines flight attendant who had called in a bomb threat so she could have the day off.

In some ways, though, Ortiz's argument almost sounded reasonable. We had been keeping our discoveries secret. That *must* be bad, right?

Until then I had simply ignored the protestations of the Internet chat group that accused us of malicious behavior, assuming that responding would simply give credence to the allegations; but as always, the Swift-boating had worked. I even started getting e-mails from real scientists asking me why we were hiding things.

I finally had to respond. I stayed up late that night and wrote a long post on my website, which ended up being reproduced around the world. I acknowledged the accusations that we hide discoveries and harm science, and then I wrote:

> One of the things that is so strange about these allegations is that they should also be made of every single scientific result that is published in every single reputable scientific journal. In all such cases, scientists make discoveries, they verify their discoveries, they carefully document their discoveries, and they submit papers to scientific journals. What they *don't* do is make discoveries and immediately hold press conferences to announce them (any scientist who ever contemplates such a thing can be stopped cold in his tracks by simply whispering the phrase "cold fusion"). Good science is a careful and deliberate process. The time from discovery to announcement in a scientific paper can be a couple of years. For all of our previous discoveries, we had described the objects in scientific papers before publicly announcing the objects' existence, and the time between discovery and announcement was always less than nine months.
>
> These scientific papers are important. They allow

other astronomers to verify, confirm, and critique the analysis we have done. Sadly, because we were forced to announce Xena and Easterbunny prematurely, we hadn't completed the scientific papers describing these objects. We find this situation scientifically embarrassing and apologize to our colleagues who are reduced to learning about this new object from reading reports in the press. We are hard at work on these scientific papers, but, as we have said above, good science is a careful and deliberate process and we are not yet through with our analysis. Our intent in all cases is to go from discovery to announcement in under nine months. We think that is a pretty fast pace.

One could object to the above by noting that the *existence* of these objects is never in doubt, so why not just announce the *existence* immediately upon discovery and continue observing to learn more? This way, other astronomers could also study the new object. There are two reasons we don't do this. First, we have dedicated a substantial part of our careers to this survey *precisely* so that we can discover and have the first crack at studying the large objects in the outer solar system. The discovery itself contains little of scientific interest. Almost all of the science that we are interested in doing comes from studying the object in detail *after* discovery. Announcing the existence of the objects and letting other astronomers get the first detailed observations of these objects would ruin the entire scientific point of spending so much effort on our survey. Some have argued that doing things this way "harms science" by not letting others make observations of the objects that we find. It is difficult to understand how a nine-month delay in studying an object that no one would even know existed otherwise is in any way harmful to science!

Many other types of astronomical surveys are done for precisely the same reasons. Astronomers survey the skies looking for ever more distant galaxies. When they find one, they study it and write a scientific paper. When that paper comes out, other astronomers learn of the distant galaxy and they too can study it. Other astronomers cull large databases such as the 2MASS infrared survey to find rare objects like brown dwarves. When they find them, they study them and write a scientific paper. When the paper comes out, other astronomers learn of the brown dwarves and they study them in perhaps different ways. Still other astronomers look around nearby stars for the elusive signs of directly detectable extrasolar planets. When they find one, they study it and write a scientific paper. This is the way that the entire field of astronomy—and indeed all of science—works. It's a very effective system; people who put in the tremendous effort to find these rare objects are rewarded with getting to be the first to study them scientifically. Astronomers who are unwilling or unable to put in the effort to search for the objects still get to study them after a small delay.

There is a second reason that we don't announce objects immediately, and that is because we feel a responsibility not just to our scientific colleagues but to the public. We know that these large objects that keep being found are likely to be intensely interesting to the public, and we would like to have the story as complete as possible before making an announcement. Consider, for example, the instantaneous Ortiz et al. announcement of the existence of 2003 EL61. Headlines in places like the

BBC web site breathlessly exclaimed, "new object may be twice the size of Pluto." But even at the time we knew that 2003 EL61 had a satellite and was only 30% the mass of Pluto. We quickly got the truth out, but just barely. Sadly, other interesting aspects of 2003 EL61 also got lost in the shuffle. No one got to hear that it rotates every 4 hours, faster than anything else known in the Kuiper belt. Or how that fast rotation causes it to be shaped like a cigar. Or how we use the existence of the satellite to calculate the mass. All of these are interesting things that would have let the public learn a bit more about the mysteries of physics and of the solar system. In the press you get one chance to tell the story. In the case of the instantaneous announcement of 2003 EL61 the story was simply "there is a big object out there." We are saddened by the lost opportunity to tell a richer scientific story and to have the public listen for just one day to a tale that included a bit of astronomy, a bit of physics, and a bit of detective story.

Given that we do precisely what other astronomers do and that we are actually very prompt about making announcements, where did the crazy idea that we should be announcing objects instantly come from? Interestingly, there is one area of astronomy in which instantaneous announcement is both expected and beneficial to all. In the study of rare, quickly changing objects, such as supernovae, gamma ray bursts, comets, and near-earth asteroids, astronomers quickly disseminate their results so that as many people as possible can study the phenomenon before it disappears or changes completely. No one discovers a comet and keeps the discovery to himself to study, because by the time the study would be done the comet would be gone and no one else could study it ever again. The people initially suggesting that we were wrong not to announce our objects instantly are, for the most part, a small group of amateur astronomers who are familiar with comet and near-earth asteroid observation protocols. We can only assume that this familiarity led them to their misconceptions. Kuiper belt objects are not quickly changing phenomena. Astronomers will be intensively studying Xena for a long time to come.

We hope to discover a few more large objects in the outer solar system. When we do, we will do everything we can to learn as much as possible about them before we make their existence public, and we will try to make the announcement as complete and scientifically and publicly interesting as possible. We will take the chance—as all scientists do—that by taking the time to do the scientific job correctly someone else may beat us to the announcement, and if they do we will congratulate them heartily.

The chat group went crazy at this point, but I never read it again, and forbade anyone from relaying stories to me. For most of the next year, Ortiz was not seen or heard from at any of the various scientific conferences around the globe. I assumed—incorrectly—that we would never hear from or about him again.

In the years that have followed, I have occasionally wondered what really happened. I will never know. In his few public pronouncements, Ortiz has claimed the only thing that he could: that he legitimately discovered K40506A/2003 EL61/Santa/Haumea one day before he stumbled upon our website, and when he had announced his discovery, there was no good mechanism for mentioning that his team had accessed our database. It was a simple oversight. What if that story is true? What if I put these guys through hell over a discovery that was legitimate? How will I ever get rid of that nagging feeling that maybe they were hardworking underdogs who had made the discovery of their lives?

But but but. But if they were on the up and up, why would they hide the fact that they had accessed the data? Why wouldn't they mention it in the early days, when Ortiz and I were exchanging cordial e-mails? Sure, there was no official channel for mentioning that they had known about our database, but given that I opened a friendly back channel the day after Ortiz's announcement, might he not have mentioned it to me then?

I went back and checked the e-mails recently. It is true that Ortiz never publicly denied having used the data, even in the early days. He just never answered the question. All of the denial came from his German friend, who, I still believe, had been equally duped. I wonder if he secretly suspected that something was amiss or if he was simply as trusting and naïve as I used to be.

Chapter Eleven
PLANET OR NOT

On that Friday morning in late July, I made the instant decision to announce to the media that Xena was the tenth planet. I had been swayed in part by Diane's arguments and in part by the urgings of the media relations person to whom I spoke that morning. But even though I was a bit blindsided that morning, all spring I had been trying hard to understand how we should define the word *planet*.

I asked an old college friend with a Ph.D. in philosophy: What does a word mean when you say it?

"Words mean what people think they mean" was his smoothly philosophical reply. "So when you say 'planet' it means what you are thinking when you say it."

I probably should have known better than to ask him. I remembered that in college he had told me that he woke up every morning surprised that reality was still reality.

Still, maybe there was something to it. Maybe words do just mean what we think they mean.

Perhaps it is wrong for astronomers to attempt to redefine a word when people already know what it means when they say it. Perhaps the job of astronomers is, instead, to *discover* the definition of the word *planet* as people use it. After all, the word *planet* has been around much longer than, well, our understanding of planets.

So what do people mean when they say the word *planet*? That spring, well before anyone knew that the world was about to be handed a tenth planet, I started asking everyone I saw. The answers were diverse and, more often than not, scientifically misguided: large rocky bodies in the solar system (well, no, there are gas giants), things with moons (not Mercury or Venus!), things that are big enough to see with your eye (Uranus, Neptune, and Pluto are out), things that pull the earth around in its orbit (that's just the sun). But when I then asked people to name the planets, everyone had exactly the same answer, starting with Mercury and ending with Pluto. People who felt themselves quite up to date and informed would then explain that maybe Pluto shouldn't be called a planet, but they certainly knew that it currently was one.

So, again, I ask: What do people mean when they say the word *planet*? They mean a slew of unscientific clutter. And then they mean nine specific objects in the solar system.

I always pressed people further: How would you know if something new was a planet? The answer was always the same: If it was as big as the other planets. Or, as I interpreted it, according to my unscientific springtime poll, everything the size of Pluto and larger that orbits around the sun is a planet.

Isn't that the real definition, then? Shouldn't astronomers leave the word alone if it already has a meaning?

I remained torn. If Pluto was a planet, why were the many things just a little smaller than Pluto not considered planets? It made no scientific sense at all. Why draw such an arbitrary line right around the size of Pluto? Isn't the job of scientists to guide the public's understanding of nature rather than acquiesce to unscientific views?

In addition to everything else happening that spring, while Xena and Santa and Easterbunny were just being found and studied, and Lilah—still known as Petunia—was growing and beginning to kick inside Diane's stomach, I was teaching introductory geology at Caltech for the first time. I'm not a geologist. I've never taken a single class in geology. If you gave me a handful of different types of rocks, chances are I could identify only a small number of them. I still get confused by the meanings of *strike* and *dip*.

Luckily, most of my students didn't realize this.

I was pretty good at teaching that class, actually. The class was the equivalent of what is called a "rocks for jocks" class at many other universities, meaning that it is intended for people who won't end up majoring in geology. Caltech, though, is not known for its jocks. All of the kids in the class who don't major in geology are majoring instead in physics or biology or mathematics or engineering. I affectionately referred to the class as "earth science for eggheads."

But why was I teaching a class about which I knew nothing? One reason only: I had begged. As an astronomer who studies planets, I have ended up at Caltech not in an astronomy department but in a planetary science department. And the planetary science department is tacked on to the side of the geology department. The people I see walking around the halls and coming to my classes tend to be geologists. After having been at Caltech for almost a decade, I thought it might be time to actually learn some geology. And what better way to learn than by teaching it myself?

I had intended to spend most of the winter preparing for the class; instead, I spent it working on the newly discovered Santa and Xena. As the first class came around in April, I was barely on track with the teaching. And then we discovered Easterbunny that week.

Still, I stayed about two weeks ahead of the class, learning the material as I went along. Over the course of the term, I said only one thing that I now know to be blatantly wrong. (To anyone who took my Ge 1 class in 2005, I apologize. The mineral peridotite does *not* change into spinel as it is compressed by high pressure; its *crystal structure* collapses to one that is identical to that of spinel, but the chemical compositions of the two minerals are totally different.)

Teaching earth science to eggheads has so far been the highlight of my teaching life. The earth is a spectacular laboratory that you can get to by simply walking out the back door. The eggheads and I took trips to the local arroyo to understand debris flows in the Los Angeles mountains; we walked one mile south of Caltech to our local thrust fault; we took a bus up the east side of the Sierra Nevada, stopping to see ancient volcanic flows, now-dry Ice Age lakes, and a 50-million-year-old mountain range now buried almost to its top in debris. All the while, I tried to pull the students out of the mind-set that is all too easy to get into in the middle of a hard first year at college: Give me the information; tell me what I need to know, what's on the exam. In earth science for eggheads the message was instead: Look around you! What is happening here? Why?

Because my head was so immersed in the geological world that spring, it is perhaps not surprising that I started looking to the earth sciences for examples of the ways in which scientists were confronted with words that had previous meanings. Geologists, in fact, have had a more difficult time than astronomers on this issue. While planets are up in the sky and don't form part of most people's everyday experiences, daily life is filled with geology. People see mountains, rivers, lakes, oceans. Or should they really be called hills, streams, ponds, and seas? When is something a mountain instead of a hill? A river instead of a stream? A lake or a pond? An ocean or a sea?

Geologists have never attempted to define these things. The words simply mean what people think they mean when they say them.

I grew up on a little rise in northern Alabama called Weatherly Mountain. As a child I assumed that the word *mountain* had some sort of meaning. When we took our first family trip west and encountered the Rocky Mountains rising six thousand feet from their base, I was stunned. Our three-hundred-foot-high mountain looked to be a molehill in comparison. But still, Weatherly Mountain will always be Weatherly Mountain.

The best geological equivalent to the word *planet* is the word *continent*. What does the word *continent* mean? As far as I can tell, the definition is something like: big coherent chunk of land. How big? The only answer I could ever find was "big enough." Australia is big enough. Greenland is not.

I began to quiz people about continents as much as I quizzed them about planets. I heard all sorts of interesting theories about how the word *continent* was defined, including a few from people who knew a little geology. I was told, emphatically: A continent is any island on its own continental plate. Greenland doesn't qualify because it is on the same plate as the rest of North America and thus is not separate. I pointed out that continents have been around much longer than the plate tectonic theories of the 1970s. And I pointed out that by the "scientific" definition, we should really count the south island of New Zealand as a separate continent.

So how do we really define continents? Simply by tradition. The seven continents are the seven continents because that's what people mean when they say the word *continent*.

But even that is not entirely true. Apparently some people mean different things. As I quizzed more and more people, I learned that, for example, many Europeans do not consider Australia to be a continent. Argentinians consider North and South America a single continent (the Panama Canal is not enough of a break for them, I guess). And rational people in many places believe that Europe is considered a separate continent only because, well, that's where the people who defined the continents in the first place all came from.

Can it really be that the most important classification scheme for our understanding of landforms has no scientific basis whatsoever? Shouldn't geologists get to work defining their terms more carefully?

And yet, when geologists talk about continental crust or continental shelves they know exactly what they are referring to. They never use the word *continent* by itself unless it is just to refer to one of those landmasses that we agree are called continents.

For the public, having a handful of continents whose names everyone can remember (even when everyone doesn't always agree) is an important way to organize our understanding of the world around us. It is too difficult to make sense of the hundreds of countries on the earth without an organizing principle. The continents are a way to bring the vastness of the earth down to a human scale.

And so with planets. The planets are our way of organizing the universe beyond the earth. In fact, they are the grandest organizational scheme that most people know. Ask someone to describe what is around them and they'll describe their neighborhood. Press further and they might talk about their town and region. If you keep pressing, perhaps they will mention their country, next their continent (*that word again!*), and finally the world. But if you don't give up there and you ask for more, you will ultimately be led down the descriptive path of the solar system. You'll be told about the planets. And after the planets? What next? More often than not, you will be left with blank stares.

When people describe their neighborhoods, they don't care about the scientific meaning of the words they're using; they care about recognizable landmarks to specify the points and the boundaries of their lives. The planets are these landmarks. That is what people mean when they say the word *planet.*

Is the word *planet,* then, specific or descriptive? When people say the word *planet,* do they mean precise places—Mercury and Venus and Earth and the others—or do they mean those places and anywhere else like them?

I found history to be a useful guide. When Uranus was accidentally discovered, it was quickly accepted as a planet; Neptune, likewise. Even Pluto, whose status was at stake in all of this, was accepted into the club with only a little grumbling. Sure, it was thought to be much bigger when it was originally accepted, much more like the other planets, but with its acceptance the bar was accidentally lowered, and most people—except for me and a few other nitpicky astronomers—meant Pluto, too, when they said *planet.*

All of this planet-or-not-a-planet business would eventually be decided by the International Astronomical Union, which, by international agreement since 1919, has the right and the responsibility to make sure that everything in the sky is categorized, named, and filed in its right place. Before the IAU came along, the skies were filled with objects named by whatever system the astronomers categorizing them chose. The reddish star in the upper right of the spectacular constellation Orion is known not just by its common name Betelgeuse, which in Arabic means "armpit of the giant," but also by HD 39801, for its place in Henry Draper's catalog from the 1920s, and by many more names, including PLX 1362, PPM 148643, and my favorite, 2MASS J05551028+0724255, in other catalogs. The IAU now has procedures and policies for what to do about almost every type of discovery in the sky. A new supernova explodes? It gets a year and a letter. Supernova 1987A was the closest and brightest in living memory, and those five characters can still provoke a glowy sigh in an astronomer of a certain age. The procedure is much more systematic, though the names are not quite as evocative as giants and armpits.

Only with the solar system does the IAU require history and poetry when it comes to naming things in the sky. By IAU decree, the moons of Jupiter are to be named after the consorts (voluntary or otherwise) of Zeus, craters on Mercury are to be named after poets and artists, and features on Saturn's gigantic moon Titan (and I can only call them "features" because we have no idea what they really are) are named after mythological places in literature.

The IAU was quick to act after the discovery of the first objects out beyond Neptune in the Kuiper belt. Things way out are named after creation gods in world mythology, though as the number of objects in the Kuiper belt grew faster than new creation gods, the rule began to be applied more and more loosely. Recently someone even got away with calling something in the Kuiper belt Borasisi, which is a god from a fictional story by Kurt Vonnegut.

For all of its preparedness for endless contingencies and countercontingencies, the IAU had never actually contemplated a question that was suddenly on everyone's minds: What do you call something new that is bigger than Pluto? How do you recognize a new planet when you see it?

Like any good international organization, it knew what to do when faced with such emergencies: It needed to form a committee. Better yet, it already had one. At about the same time that Xena was discovered and astronomers (and everyone else) began to try to figure out how small the smallest planet could be, astronomers were also struggling with the question of how big the biggest planet could be. We had it easy in the solar system; it was unlikely that I would ever find anything bigger than Jupiter that needed to be categorized, but discoveries of things larger than Jupiter orbiting around distant stars were beginning to become routine. Some were about as massive as Jupiter or only a little more so. They were clearly planets. Some were only somewhat less massive than the sun. Those were clearly stars. Some were in between. What to do? The IAU formed a committee to decide. This same committee was now charged with figuring out the small end of planets, too.

When the reporters called to talk about the discovery of Xena, they wanted to know where it was, how we had found it, and how big it was. We didn't yet know the size for sure but thought it could be as much as half again the size of Pluto. And then they asked: When is the IAU going to make a decision about planets?

"I hope they'll decide before my daughter starts crawling," I joked when Lilah was three weeks old.

When a bevy of new reporters called after the story about the Spanish snooping broke, they asked what was going to happen to the Spaniards, how the conflict was going to be resolved, and how this might change the way astronomers interacted and protected their data. I explained that many scientists worldwide suddenly realized that they, too, could be vulnerable to unintended snooping, and many were scrambling to find solutions. And then they asked: When is the IAU going to make a decision about planets?

"I hope they'll decide before my daughter learns to stand," I joked a few months later, since Lilah was already crawling by that point.

When we discovered that Xena was not alone at the edge of the solar system, that it had a tiny moon going around it, reporters called again and wanted to know how the moon got there, what it looked like, and what we were planning to call it (Gabrielle, of course, after Xena's spunky TV sidekick). And then they asked: When is the IAU going to make a decision about planets?

"I hope they'll decide before my daughter says her first words," I joked the next winter, since by then Lilah was already standing and taking increasingly assured strolls around the periphery of the room.

In the spring, when we finally were able to use the Hubble Space Telescope to figure out just how big Xena really was, reporters called again and wanted to know what Xena was made of, how it had gotten so big, and how much larger than Pluto it was (only 5 percent, it seemed, which was uncomfortably close to not even being bigger than Pluto at all, particularly when you include approximately 4 percent uncertainty in the measurement). And then they asked: When is the IAU going to make a decision about planets?

"I hope they'll decide before my daughter goes to college and takes an astronomy class," I finally reverted to joking when it became clear that no decision was going to be made anytime soon.

People kept asking me when the IAU was going to make a decision because they thought I should know. But I didn't know anything. During this entire period, no one officially connected with the decision making—and I didn't even know who that might be—ever once contacted me to ask a question or to tell me what was going on. I assumed that I was going to wake up one morning, open the *Los Angeles Times,* and see that I was suddenly the official discoverer of a planet. Or that there were only eight planets. Or that I had discovered many planets. Or that I had discovered the only thing in the solar system larger than a planet that wasn't a planet.

In the face of this uncertainty, I figured it best to be prepared for all options. I called up the person in media relations at Caltech who had—months earlier now—pressed me to decide whether or not to call Xena a planet in that original press release. I told him that we needed to prepare another press release, this time for the IAU decision.

"Great!" he said. "What did they decide to do?"

"Well, actually, they haven't decided anything yet."

"But they're deciding soon, right?"

"Well, actually, I have no way of knowing what they are doing. They might decide tomorrow, and they might decide a decade from now."

"So . . ." He paused. "What are we going to say in a press release?"

I knew that I wanted to have the opportunity to tell the full scientific story to the public. I had missed that chance, I felt, back in the original hurried rush when we had to make the announcement right away. I wanted the beauty and subtlety and essential order of the solar system to be at the center of the discussion after a decision was finally made. I cared less what the IAU decided—within limits, of course—than that the science got explained correctly.

"We're going to write four different press releases," I explained.

Ten planets made sense if you wanted your planets to have more emotional resonance than scientific significance. We very quickly wrote that press release, hailing Xena as the tenth planet. It made me proud to think of my tenth planet, but even from early on I admit that it also made me feel a little fraudulent. The discovery of Uranus was a big deal, and that of Neptune was amazing. But Xena? Little Xena? The tenth planet? Still: I channeled my inner geologist. If it mattered emotionally, that was all that counted. I was ready.

Scientifically, I agreed much more strongly with our second press release, explaining why there were only eight planets. Eight planets made sense if you were a scientific historian and realized that 150 years ago people had already decided to divide the solar system's objects into big planets and small asteroids and that Pluto—and now Xena, too—thoroughly fit into the category of the small objects. I liked to think of this one as what people would mean when they said the word *planet* if they really understood the solar system. We hailed the bravery of astronomers for taking a scientifically sound stance in the face of what would certainly be considerable opposition. Even though scientifically I agreed much more strongly with this press release, I was glad we wrote it quickly; it seemed quite clear to me that astronomers would never have the audacity to actually get rid of everyone's favorite runt planet. Still, to be safe, I thought it best that we have a press release ready. Having the discoverer of the onetime tenth planet agree that it should not be a planet seemed like a powerful line of argument.

Our third press release was for the possibility that the IAU's decision would be to simply keep nine planets. Keeping Pluto in the planet club but refusing to allow in bigger newcomers didn't make sense at all. Yet it did seem like an option that might be on the table, since I had heard a few people say, "Why do we need to change anything when we have nine perfectly good planets?" Our press release said that nine planets was a pretty dumb decision.

The last press release considered a more extreme possibility: that the IAU would stretch the definition of the word *planet* so far that there would suddenly be two hundred planets. A small but extremely vocal group of astronomers had been pushing for a while to thoroughly transform the meaning of the word *planet.* Unlike the ten-planet approach, which was an attempt to understand what people meant when they said the word *planet,* or the eight-planet approach, which was an attempt to discern what people would mean if they had all of their facts straight, or even the nine-planet approach, which was to stick to literally what people meant when they said the word *planet* (the nine planets and nothing else!), the two-hundred-planet approach was an attempt to legislate an entirely new and never before anticipated meaning for the word *planet.* The word was to mean, essentially, "anything in orbit around the sun that is big enough to be round."

Why round? It is not simply that astronomers are enamored of that particular shape (though, really, why wouldn't they be?). It is that that particular shape tells us something. If you throw a boulder into space, it will retain whatever irregular shape it originally had. If you throw a hundred boulders into space together, they might stick to one another due to the tiny amount of gravitational pull that each boulder generates, but they could still have almost any shape you might imagine. But if you put enough boulders up into space together, a fantastic thing will happen: The cumulative gravitational pull of all those boulders will take over. The boulders will pull together and crush and smash one another until you can no longer discern what shapes they had to begin with; instead, they will form a beautiful, simple sphere. Finding something spherical in space indicates that you have found a place where gravity has taken over. I am pretty certain that at no time in the previous several-thousand-year history of the word *planet* did anybody say "planet" and really mean "things that are round due to self-gravity." It was a new definition by simple fiat. And it would lead to something like two hundred new planets, most of them out in the Kuiper belt.

I explained all of this to the person writing the press releases.

"Why bother writing this one up? It sounds crazy. No one would decide this, would they?"

Well, yes. I actually thought this was the most likely decision that the International Astronomical Union would make—if it was ever going to make a decision at all.

"But why would astronomers do such a crazy thing?" he wanted to know.

Desperation was all I could answer. Desperation.

As radical as the new definition was, it was the only one on the table that both smelled scientific and also retained Pluto as a planet. I could imagine that it would be hard for a scientific committee to conclude that the definition didn't need to have a strong scientific basis (the nine- or ten-planet approach). I could imagine that an astronomical committee would be loath to provoke what would certainly be a huge public outcry if it kicked out Pluto (the eight-planet approach). So although the two-hundred-planet approach was the most radical, it had the cover of appearing the most conservative. I could just see it passing.

I didn't like the definition, but I could live with it. The good news, for me, was that if this new definition was announced, I would have discovered more planets than anyone else in human history. Not just Xena, Easterbunny, Santa, Sedna, and Quaoar, but dozens more. The bad news was that I couldn't remember most of their names.

. . .

The first anniversary of the discovery of Xena came and went with no hint of what was happening at the IAU. But it was okay. I was busy. Chad and David and I, now joined by some of my students and outside colleagues, wrote scientific papers about the size of Xena, the discovery of Gabrielle (the moon of Xena), the discovery of a thoroughly unexpected second moon of Santa, and the slab of frozen methane covering the surface of Easterbunny; and we still had much, much more to do. There were press releases to develop, talks to give around the country, interviews on TV and radio. But when I think of this time period, I have a hard time remembering almost any of it. What I really remember is Lilah and the moon.

As with any overeducated first-time parents, we were fascinated with understanding Lilah and what she was thinking and doing and understanding. I began reading scientific books on early childhood development, not as a way of pushing Lilah along faster or making sure she was okay, but simply because it was, at the time, the single most fascinating thing I could imagine. I read studies of the development of facial recognition and motor skill control, but what I found the most interesting of all were studies of language development. It seemed so hard for me to imagine that this little baby, being carried around in a bundle in my arms, would someday be sitting in a chair next to me having a conversation.

Diane and I often joke about parents who think that everything their children do is exceptional. Intellectually, we always understood that Lilah would likely be good at some things, not as good at other things. Exceptional is a pretty high bar. But reading these books about early childhood and watching Lilah develop, I finally understood. She *is* exceptional, because early childhood development is about the most exceptional thing that takes place in the universe. Stars, planets, galaxies, quasars are all incredible and fascinating things, with behaviors and properties that we will be uncovering for years and years, but none of them is as thoroughly astounding as the development of thought, the development of language. Who would *not* believe that their child is exceptional? All children are, compared to the remainder of the silent universe around them.

Emily Schaller, my Ph.D. student, who was always willing to try to engage my obsessions, handed me a book one day on how to teach your baby a primitive type of sign language. The thought was that children are ready to communicate before they have the vocal motor skills for speech. But they can use their hands and arms and fingers to tell you about the world around them.

Lilah's first sign was, not surprisingly, *cat* (rake two fingers across your face as if to draw whiskers). The two cats—originally Diane's, now a joint venture—who lived with us barely tolerated the loud newcomer to the house. But they eventually got used to her and realized that she posed no harm, so they would lie close to take advantage of me or Diane being immobilized while holding a sleepy baby and having a free hand available for an ear scratch. Then Lilah learned to roll over. The cats scattered, to eventually return when they realized she was still *mostly* immobile. Then Lilah started to crawl, and that started the years of Lilah chasing after the cats, the cats slinking away, always out of reach. To Lilah, the cats must have been like the end of the rainbow: always in sight, always just out of reach, and gone when you get there. Her first efforts to communicate with the external world were targeted directly at them. They, sadly, never returned the favor.

After *cat*, Lilah next learned *flower*. Flowers (scrunch up nose as if sniffing) were everywhere, first only outside on plants, but soon she generalized to flowers on her clothes or her shoes, or in pictures in books and magazines. I wanted to hook up wires and do experiments and comparisons and studies to understand it all.

"You want to do what?" Diane would say.

But, really, who wouldn't? In our own house the most extraordinary thing in the universe was taking place, and it was passing by unexamined, unstudied.

"There will be no Lilah experiments," Diane declared.

I know, I know. I wouldn't *really* do it. I didn't *really* want to hook up those wires. Mostly I just wanted to hold Lilah tight as she made signs to the world around her, and I wanted to tell her: You are the most extraordinary thing in the universe.

We had recently bought a new house. For the first few years of our marriage and the first six months of Lilah's life, we had lived in a diminutive Spanish-style bungalow in a typically densely packed part of suburban Pasadena that I had bought years earlier. I loved my little bungalow. It was the house where I first cooked dinner for Diane. When Diane had moved in, I had warned her: I love this place and never want to move.

But the house had almost no sky.

I biked home at night through lit streets, car headlights glaring everywhere. I thought back to the days of living in the cabin and walking the trail by the light of the moon or stars; I thought back even earlier to the days of living on a tiny sailboat in the San Francisco Bay and staring up at the whole sky before finally closing the hatch for the night. From my bungalow, I could sit in the hot tub in the backyard and look up and see slivers of sky.

Sometimes I could see the Big Dipper, sometimes Cassiopeia. But in my tiny night-sky universe, I never once saw a planet.

When Diane suggested that we probably needed to move to a bigger house to fit our now-expanded family, I reluctantly agreed. Maybe it was time. I grudgingly went to look at a few places with her. Nothing felt as perfect as our happy little bungalow. Then one day, with no expectations, we stumbled onto a house perched on top of a massive one-hundred-thousand-year-old landslide. Almost nobody knew it was a landslide, but I had made my geology students write about it nearly a year earlier. How could I not have fallen in love with the house? We bought it three days later and moved in the following month.

Living on a landslide has its advantages. We have a steep canyon in our backyard, because canyons are easily formed in rubble. We have landscaping boulders of every conceivable size and composition, and if we ever have the need for more, we just dig a foot underground to see what else the landslide brought down. The landslide makes a minor wildlife corridor, so we have an abundance of birds, an occasional bobcat, and even once a black bear.

For me, though, the best benefit of perching on the tip of the tongue of a landslide with the mountains rising to the north of you is that you have an uninterrupted view to the south. And if you go outside at night and look to the south, you get to see the spectacular constellations. You get to see Orion and Taurus and Scorpius. You get to see the blue of Sirius, the red of Betelgeuse. And best of all, you get to see the planets.

Lilah and I have spent the years since we've moved to our new house tracking Jupiter and Saturn across the sky, watching Venus set into the Pacific Ocean, seeing how red Mars looks in comparison to the pale stars. But more than anything else, we've watched the moon.

When we first moved into the new house, Lilah was still learning new word signs. One of her favorite combinations translated something like this: There is a light; turn it on (hand held overhead, first balled momentarily, fingers suddenly flying open to show you what to do). If you complied, she would even say "thank you" (finger tapping heart).

One spring night, the nine-month-old Lilah and I sat outside wrapped in blankets staring up at the moon, which was nearing full. A rollicking storm had been through for the past few days, revealing to us all of the places where our new house flooded. But the rain had stopped, and between the thick black clouds covering about half the sky we could see the brightest stars and the bright full moon, which had found itself in a large hole in the clouds and was shining down on the still wet and now sparkling nighttime landscape. I told Lilah about night and the moon and the rain. We heard coyotes across the canyon, and I told her about them, too (and about why the kitties were now going to be solely indoor kitties).

And then the moon ducked behind one of the thick clouds, and everything got dark.

Lilah looked around, looked up to where the moon used to be, and looked at me. Then she held her fist up in the air and flung her fingers open. She looked at me expectantly.

The cloud passed. The moon came back out and once again brightened the landscape.

Lilah smiled at me and tapped her heart.

. . .

I have an extremely vivid memory of the day that summer, a few weeks before her first birthday, when Lilah really learned to walk. She had previously taken a few halting steps before tumbling, or she had scooted along while holding a wall, but one day she instantly went from easily manageable (she would not have gone too far if I had looked away for sixty seconds) to fast, unpredictable, and apt to disappear in seconds. A day earlier, while trying to throw a friend in the swimming pool, I had broken my ankle, and I was now in a cast and on crutches. The fact that I had to take a few steps from the kitchen counter to the refrigerator was something I now mourned. But the worst thing was that Lilah was suddenly up and running just as I was slow on my crutches. The only way I could keep up was to crawl. So, that day, Lilah and I traded. She now walked. I now crawled. I had many theories about the precise symbolism of that transition, and all of the theories were ominous.

Lilah made up a sign-language symbol for me walking on crutches (two hands held in front of her, pointer fingers moving up and down), which I count as the moment when she first learned to mock me. I was, perhaps unsurprisingly, enamored of the mocking.

Six days later, still on crutches, I headed to Italy to give a talk at an international conference on the Kuiper belt. We talked about the formation of the Kuiper belt, the surfaces and atmospheres of the objects there, and what they might be made of, but the question of what to call them never came up. But at night, when we went to little cafés (the closest of which was precisely 1,032 crutch-steps away, which felt to me like the distance from Earth to Sedna) to drink prosecco and watch the World Cup soccer games, everyone wanted to speculate about Pluto and Xena and planets. I tried out my arguments about ten planets and about continents. The scientists balked. They didn't like the idea that the definition of *planet* would include no science.

"So you think everything round should be a planet? You think that there should be two hundred planets?" I asked, assuming that that was going to be the obvious response.

"Of course not!" they responded. Wasn't it obvious that there were only eight planets?

I thought the other astronomers were being naïve. It's easy to sit inside the scientific bubble and make pronouncements, but they were forgetting how much of an impact this decision was going to have on the outside world. No one was going to let Pluto be killed, were they? But still, it was interesting. Within the field of people who studied the Kuiper belt for a living—people who had devoted their careers to the outer solar system and its many, many denizens—it was almost not worth having the conversation. Of *course* Xena was not a planet. And Pluto likewise. Hadn't we settled that question 150 years ago when the asteroids became asteroids?

Naïve, I thought. I remembered back to the days when I used to think the exact same thing. Wouldn't it be nice to just think about science and not worry about its impact on culture? Wouldn't it be nice to be able to just say the thing that makes the most sense?

A week after I got home, my phone rang, and out of the blue I was told by a member of a previously unknown IAU committee (the third planet committee? The fifth committee? I couldn't keep track) that Xena was to be a planet.

He couldn't reveal the details of the decision on the definition of the word *planet*, but he wanted to prepare me for the onslaught of publicity that would surely follow. As I was the only living discoverer of a planet, he thought it best that I stay humble.

Humble? I thought, and chuckled to myself. My one-year-old daughter had recently learned to mock me in a sign language she had made up herself.

While he had not meant to reveal details to me, he already had. The "only living discoverer" could mean only one thing. If the IAU was going to pick the two-hundred-planet definition, there would have been perhaps a dozen living planet discoverers. If there was only one, it was clear that the IAU had decided on the ten-planet definition that I had come to terms with myself. Xena was to top off an elite list.

"Do you think the rest of the astronomers will go along with this?" I asked.

I was quickly assured that they would. "I've had a lot of conversations in the past few days. This is going to sail through the vote process."

. . .

I went home that night and told Diane. We opened a bottle of champagne and drank to the amazing fact that I had discovered a planet. *A planet. I had discovered a planet!* After all of this time, Xena was officially going to be a planet, and I was officially going to be the only person alive who had found a planet.

Just then, Lilah walked around the corner from where she had been playing, saw the crutches under my arms, and immediately stuck out her hands and waggled her pointer fingers.

Okay, so I was slow, and I still had to crawl to be as fast as my one-year-old daughter. But I had found a planet. No one could take that away from me.

Chapter Twelve
MEAN VERY EVIL MEN

Living on the outskirts of Los Angeles with a clean sweep of the sky to the south of us, we have a very nice view of the standard flight path for arrivals and departures at LAX. Those things moving through the daytime sky that turned into lights brighter even than the stars at night held a special fascination for Lilah. Her sign-language symbol for airplane (arm held aloft with hand parallel to the ground) got much use. First it was just for those little moving dots in the sky, then it was for pictures of airplanes in books, and then, one very exciting day when Lilah was thirteen months old, it was for an airplane that she actually got to fly in. I spent the whole morning trying to prepare her for the mental transition:

"Look! Airplanes in the sky!" I said, as we got close to LAX.

"Look! The airplanes are driving around on the ground!" as we were moving through the airport.

"Look! That is the tunnel people take to get *on* to the airplane!" as we were at the gate.

"Look! We're now inside!" as we sat down.

For what I had assumed would be a difficult cognitive leap, Lilah took it all in stride. Of course we're inside the airplane and now flying in the sky, Daddy. What else would we be doing?

We were taking our first family vacation, two weeks on Orcas Island, the largest of the San Juan Islands, northwest of Seattle. Diane had lived on Orcas Island for her high school years, and her mother still lived there. It was the first place Diane and I had ever gone together on an official vacation (Hawaii together? That was just work). And now we were bringing our daughter there. Diane has an affinity for the annual Library Fair, where used books get sold on a Saturday morning in August at the otherwise sleepy farmers' market. For reasons that I, as a non–native islander, can never fully understand, the Library Fair is an unspoken homecoming day; previous island inhabitants show up, as if by accident, to stroll around on Saturday morning, peruse the used books, and snack on barbecued oysters.

I love the Library Fair and—out of sheer exuberance— always buy used books that I would not otherwise even glance at. And then I sit on Diane's mother's porch on a midsummer twilight lasting until well past ten o'clock and read.

But during this very first family vacation with Lilah, I had no opportunity to relax and read on the porch. My vacation coincided with the once-every-three-years meeting of the International Astronomical Union, this year in Prague. And at this meeting, for the first time in history, they were finally going to vote on the definition of the word *planet*.

Why wasn't I there? Why was I vacationing half a world away from where the astronomical action was?

It's a good question.

There are four answers to that question. First, I love the Library Fair. Second, it was our first family vacation. Third, no one—not even the astronomers who found themselves in Prague—had been warned that the vote about planets was imminent. I admit, had I known ahead of time that this vote was going to take place, I might have felt obligated to be there rather than to abscond to a small island in the Pacific Northwest. Luckily, I didn't know. Fourth, and probably most important, I am not a member of the International Astronomical Union. I would have been ineligible to vote. I'm embarrassed to admit that I can't get myself to fill out the paperwork to join. It's all because of Question 12 on the form. After the form asks for your academic qualifications and awards, both of which I can handle, it then quizzes you about your miscellaneous special distinctions. And then I stop. Well, I think, I'm no William Herschel (the discoverer of Uranus, which is indisputably a planet). I'm no Adams or Leverrier or even Johann Galle. (Adams and Leverrier predicted the existence of Neptune, and Galle confirmed it.) Really, I wonder, do I have *any* special distinctions? And each time I get to that spot in the application, I have to stop and put down my pen. There was no reason to bother going to the International Astronomical Union meeting in Prague because I would not have been let in the doors.

. . .

I was sitting in Diane's mother's house on Orcas, watching the sailboats navigating Westsound out the window, when an e-mail arrived from the other side of the world telling me the details of what, precisely, the IAU was going to vote on. I read it to Diane in my excitement.

"Diane, Diane, it says right here that the planets are to include the big eight, of course, and then also Pluto and 2003 UB313—that's Xena—and wait a second, there are a few more."

I was confused. Xena and Pluto were nine and ten. But there was also to be Ceres—the asteroid discovered in 1801 that was declared not-a-planet sometime around 1850. And in a surprise that I had never anticipated, Charon, the moon of Pluto, which was about half the size of Pluto, was to be number twelve. Twelve planets. Not eight, nine, or ten, or even two hundred, which I would have understood. And Charon? The e-mail didn't make any sense. I didn't recall any discussion in which naming Pluto's moon a planet had ever come up before. What was the committee thinking? Who in their right mind would declare Charon a planet?

I reread the e-mail carefully. The committee, which had met in secret, was adhering to the notion that "all round things are planets," which I had thought was a bad idea to begin with but which I at least understood and could support as a scientifically rational and consistent definition, even if a poorly chosen one. If the assembled body of astronomers thought that that was the right way to define the word *planet,* I would be disappointed personally, but I would get over it. After all, I would still get a few planets out of it.

The secret committee had its reasons, which were passionately stated. First: The word *planet* should have a scientific basis. Who was I to argue against that point? I had been willing to go along with a cultural definition instead of a scientific one, but if the astronomers were going to insist on science, I could hardly say no. And second, they suggested that in deciding whether an object is a planet or not, you should be able to tell just by looking at it—in other words, you shouldn't have to know anything about where it is and what it is doing and what else is around it. The committee didn't buy the idea that planets should be the small number of unique important dominant objects in the solar system.

And then they discussed the newly proposed twelfth planet, Charon.

Charon is the biggest of Pluto's three moons. It was discovered, accidentally, in 1978 by James Christy, an astronomer at the United States Naval Observatory who was examining old photographs of Pluto and noticed a slight bulge coming and going, first on one side and then the other. Though Charon is smaller than our moon, than four of Jupiter's moons, than one of Saturn's moons, and than one of Neptune's moons, making it only the eighth largest moon in the solar system, it is big proportionally to Pluto. And because it is big in proportion to the planet around which it orbits, it alone of all of the moons in the solar system deserved to be a planet.

What?

In the proposal from the committee, Charon was considered to be a planet for two reasons. First, it was big enough to be round, which was in itself a good enough reason to be considered a planet if you're inclined to think of planets that way. But there are many round objects in the solar system that no one considers a planet. My nemesis the moon, for example. In fact, the proposal from the committee specifically excluded moons from being called planets. But it made a special exception for Charon—smaller than our own moon by a factor of about sixty—for one reason: Pluto and Charon go around a center of mass that is a bit outside Pluto.

Here a bit of quick physics is necessary (and I would like to point out that the fact that we need a physics lesson to explain the proposed definition of the word *planet* is already a bad sign). Whenever an object orbits another object (the moon about the earth, the earth about the sun, for example), it is not that the bigger object is stationary while the smaller object goes in circles.

Instead, both objects go in circles around what is called the center of mass. You can find the center of mass of the earth and moon, for example, by finding a really large seesaw, putting the earth on one end and the moon, which has only 1 percent the mass of the earth, on the other end, and trying to make them balance. In the case of the earth and moon, you would have to move the pivot point to a location about a quarter of the way inside the earth. The seesaw is now balanced, and you have found the center of mass. In the twenty-nine days it takes for the moon to go in its big circle around the earth, the earth, too, in addition to traveling around the sun, goes in a tiny circle that is smaller than the earth itself. Rather than the moon circling the earth, both objects really circle around the point inside the earth that is the common center of mass.

There is nothing particularly special about the location of the center of mass. If you were to find yourself at the precise spot that is the center of mass of the earth-moon system, the only thing unusual that you would notice is that there would be one thousand miles of rock on top of your head.

Pluto is only about twice the size of Charon, so if you put Pluto and Charon on the cosmic seesaw you would find that the balance point is a little bit outside Pluto, rather than inside it. Again, there is nothing particularly special going on there. If you were to find yourself at that precise spot, you would only notice that you were very, very cold and could no longer breathe.

According to the IAU proposal, though, the obscure fact that the center of mass of the Pluto-Charon system sat a bit outside Pluto rather than a bit inside it made all the difference. It suddenly turned Charon into a full-fledged planet, and Pluto-Charon into the solar system's only double planet. Pluto lovers everywhere would be thrilled. Pluto's status was about to change from imperiled to wildly distinctive. It would suddenly be the only place in the solar system you could go and get two planets for the price of one.

Except, of course, that the proposed definition was crazy. The members of committee first argued that only the object itself, and nothing else nearby, should be considered in determining whether the object was or was not a planet. Then they changed their minds and argued that satellites, though round, were not planets, because they were in orbit around larger round things instead of the sun. And then they changed their minds again and said that Charon, though small in comparison to other satellites in the solar system, *was* a planet because the common center of mass of the Pluto-Charon system was outside Pluto rather than inside it, so that, technically, Charon orbits an empty spot in space rather than Pluto. Because it doesn't orbit a planet, it was therefore not—by this argument—a satellite.

So here is how you tell, in the committee's opinion, that something in the vicinity of the sun is a planet. Look at it and see if it is round. If it is, then it might be a planet. Next, check to see if it orbits around something else instead of the sun. If it does, then it's probably just a moon and not a planet. But before you know for sure, calculate the center of mass (if you even know the masses of the bodies in question, which you usually don't) and see if it is inside or outside the larger body. Then you know. It's all quite simple.

While the inclusion of Charon was the most jarring aspect of the proposal, there was one other oddity that I couldn't make sense of. The committee said that all round things were planets (except for moons, which weren't, except for Charon, which was). I had estimated that about two hundred objects in the solar system would fit that criterion, but the IAU had done its own estimate and come up with its own number: twelve.

Why would Charon and the asteroid Ceres be added, but not the dozen known Kuiper belt objects that were larger than Ceres? And the hundreds that were smaller but almost certainly round? It was as if the International Arboreal Union were to tell you that all things with trunks and bark and branches and leaves were to be called trees, but then it told you that the only trees were oak trees, maple trees, and elm trees. You would be right to ask: How can you make a very precise definition of *tree* and then claim that things that very precisely fit your definition are not, in fact, trees?

Why would the International Astronomical Union do such a thing? I have a theory that I strongly believe to be true, but which is strongly denied by everyone I've talked to who might have more intimate knowledge of how the decisions were made. My theory is that the IAU decided that keeping Pluto as a planet and adding three new planets—Xena, Charon, and Ceres— would seem like a minor change to the order of things. It knew that after the newspapers declared that the solar system now had twelve planets and it proudly exclaimed that its new definition was the first true scientific definition, the pro-Pluto crowds would be satisfied and no one would be terribly startled. Three new planets? Yeah, that happens every century or so. No need to get alarmed. Who could complain? It wouldn't elicit anything like the reaction people would have to the headline "Solar System to Have 200 Planets!" Given the choice between scientific rigor that might cause protests and a scientific whitewash to conceal reality, the IAU chose the latter. The first scientific definition of the word *planet* was afraid of its own scientific shadow.

From my increasingly stressful vacation spot on Orcas Island, I got ahold of the committee member with whom I had originally spoken, who was in Prague to present the committee report in the next day or two. I told him I thought the committee recommendation was a mess. How could Charon be a planet? How could it say there were only twelve round things? It made no sense.

He calmly explained the committee's reasoning and said that he would make sure that in the press release and the press conference it would be very clear that many, many more objects were on the way to being included as planets. And, he mentioned again, he had already talked to many of the astronomers in Prague, and there was nearly universal support for the new definition.

There was clearly nothing I could do. I was on a remote island on the wrong continent; it was impossible for me to have any influence over what was happening in Prague. And in Prague the very next day, they were going to declare that I was one of only seven people in human history who had ever discovered a new planet in the solar system. Who was I to complain?

That night, long after Lilah was asleep and Diane had crawled into bed, I walked (limped, really, but I was now in a walking cast, at least) down to the rocky shore. I could see north across the strait to the islands off the coast of Canada. I could see the deep twilight still casting its final red rays on one of the triangular volcanic peaks over toward the mainland. I turned around to look back at the island, back to the south, but my view of the southern sky was blocked by madrone trees growing close to the water. Farther down the beach, some rocks jutted out into the strait; I'd have a view from there. I hobbled down to the rocks and slowly made my way out to the point. From here I could see the unobstructed southern sky. Low to the south, masquerading as the brightest star around, was Jupiter, undisputed king of the planets.

I sat on the rock and watched the sky and looked at Jupiter. Who was it who first noticed that Jupiter moves? You can sit and stare at it all night long and not tell a thing. You can come back the next night, and unless you're looking very, very closely, you will still probably not notice anything different. But Jupiter moves. It's a wanderer. A planet.

I know we're past the point that when people say the word *planet* they mean an object that wanders across the sky. In fact, we're so far past that point that most people never even realize that the planets really are up there wandering night after night. Planets are, to most people these days, pictures from spacecraft, drawings on a lunch box, models in a museum. Meanings can change. After tonight the word *planet* would change again, adding to the pantheon a little point of light moving through the sky that almost no one other than I had even seen. But it was real enough to me. At any point, day or night, winter or summer, if you walked up to me unannounced and said, "Quick! Where is Xena?" I could point an outstretched finger somewhere in space and locate it, with an error of about a hand's width. If you asked me, "How big is Xena?" I would point at the moon and say, "Imagine a frosty world about half that size." If you asked me what it would be like to walk on the surface of Xena, I would ask you to image walking on a frozen lake in the dark of the new moon. That was Xena. My tiny, frozen, nearly invisibly lovely planet. I looked off to the east, where Xena was just about to rise above Mount Constitution, and thought: So be it. I was ready for the next day.

I stared back at Jupiter, wishing I had thought to bring along binoculars so that I could pick out its miniature solar system of icy moons circling it. I tried to pretend I could tell that Jupiter was moving across the sky. The earth rotated. The stars moved toward the west.

I couldn't accept it.

The solar system does not consist of twelve planets and then everything else. That is simply a fundamentally incorrect description of it. And the next day in Prague, astronomers were going to stand up and encourage the world to think of the solar system incorrectly. As someone who spends much of my life trying to be not just a scientist but an educator, trying to explain the universe and show the excitement without resorting to science fiction or trivial simplification, the idea that astronomers would actively encourage people to have the wrong view of the solar system seemed almost criminal. The idea that I was going to, overnight, become one of the most famous astronomers in the world on account of this criminal activity made me an passive accomplice. I had to do something to stop it.

I hobbled back from the rocky beach up to the house. I woke Diane and told her that when the press called tomorrow I was going to have to tell them why the new proposed definition of *planet* was no good and why, in the end, it made sense all along for there to be just eight planets. I told her that I was going to have to kill Pluto and that Xena would go down as necessary and important collateral damage.

All along, Diane had been more practical than I was. "Just let it be a planet," she would say. "Try not to worry about it so much," she had told me all year. "Relax" was her usual advice.

But this time, when I told her that I couldn't support Xena's becoming a planet, Diane simply said, "Of course not, sweetie. You always needed to do what's right." And then she gave me her usual advice: "Relax."

I did not sleep well that night.

The next morning, I went to the village of Eastsound, where I knew I could get freshly brewed coffee and a freshly flown-in newspaper. On the front page, a headline screamed, "Three New Planets Added to Solar System." A beautifully prepared graphic—courtesy of the IAU—showed the new solar system with the twelve planets all in place. The article prominently featured quotes of mine from previous interviews about the new planet Xena.

I felt sick to my stomach.

This was it. Astronomers had taken a beautiful and subtle solar system and turned it into a cartoon. And the cartoon was wrong.

I went back to the house and called the people at media relations at Caltech and told them where to find me. I hung up the phone and waited for two minutes before it rang.

I spent most of the next twelve hours, and indeed most of the next week, on the phone talking to the press about the solar system, planets, and why the IAU's proposed definition was fatally flawed, and explaining why Pluto—and Xena—should really not be considered planets.

At first the reporters were shocked. They were calling to get quotes from the most newly minted planet discoverer about how fabulous all of this was. Instead I was telling them that everything they had heard from the IAU the day before made no sense. Suddenly there was a controversy. My phone kept ringing.

Lilah developed a new sign, which either meant "Daddy" or simply meant "phone," I could never tell. Whenever she saw an object of the right size, she would pick it up and immediately hold it to her ear and then point at me.

Astronomers around the world picked up on some of the silly implications of making Charon a planet simply by virtue of the location of the center of mass of the orbit. In the middle of one phone interview, it suddenly occurred to me that the center of mass of the sun and Jupiter lies outside the sun, so by IAU logic, Jupiter should not be considered a planet since it doesn't really go around the sun. Another astronomer sent an e-mail showing that if a massive moon were on an elongated orbit, the center of mass could be inside the planet during part of its orbit but outside the planet during other parts of its orbit, meaning that, according to the IAU, that moon would switch back and forth between being a planet to being a nonplanet during the course of its orbit. And a few days later, courtesy of a fabulous press release by Greg Laughlin, an astronomer at the University of California Santa Cruz, the newspapers explained that because our moon is slowly moving outward, away from the earth, in a billion years or so it will have moved so far away that the center of mass of the earth-moon system will lie outside the earth. Suddenly: *boom!* The moon will officially be a planet. It would be a day to celebrate.

I wasn't in Prague, so someone else will have to tell the details of what actually happened there. What I do know is this: Astronomers there, who I had been told were going to go along sheepishly with this mess of a proposal, revolted.

The revolting astronomers, who grew to be a sizable fraction of the astronomers present, made it quite firmly known that they would not support the secret committee proposal. The only proposal they would support would be one where Pluto was put in its logical—rather than emotional—place. Pluto, Charon, Ceres, and my own Xena would all have to go. The press, and indeed the astronomers in Prague themselves, were quite amused by the fact that one of the most vocal supporters of demoting Pluto, Charon, Ceres, and Xena was the guy who had the most to personally gain from Xena being a planet: me.

My phone calls with the press and conspiratorial e-mails with astronomers in Prague continued for most of two weeks, first from Orcas and then from Pasadena, after we returned home from our vacation. Everything was building toward the final afternoon of the final day of the IAU meeting, when the deciding vote on the definition of the word *planet* would finally be held.

The vote was going to be broadcast live around the world, and I was going to host a packed media event to watch it, even though afternoon in Prague was before dawn in Pasadena. By 5:00 a.m. the news crews and I were setting up in a room usually taken up by press conferences about the latest southern California earthquake. The vote on the resolution that could completely change the way people looked at the solar system was slated to take place in under an hour. That morning, the astronomers in Prague had awakened to read the final wording of the resolution to be voted upon. And the wording mattered. Cosmic distrust in Prague was running so high at this point that many assumed that the clearly pro-Pluto secret committee would attempt to subvert the clearly anti-Pluto-as-a-planet majority by sneaking in wording that would keep Pluto no matter what the vote.

Sitting at a desk in the earthquake room in Pasadena in front of the still assembling press, I projected onto a large screen a just-posted copy of the precise wordings of the resolutions, which I had found at the meeting website. Along with the news crews and, by now, a growing crowd of interested onlookers from the Caltech community, I read, for the first time:

Resolution 1: Precession Theory and Definition of the Ecliptic

I only then realized that there was more on the agenda than Pluto and that this morning might be much longer than anticipated. While I know what precession theory is, and I even knew the definition of the ecliptic, I wasn't even slightly interested in knowing the precise definition being proposed here—and neither were most, if not all, of the astronomers in Prague.

Resolution 2: Supplement to the IAU 2000 Resolutions on reference systems

Yawn.

Resolution 3: Re-definition of the Barycentric Dynamical Time, TDB

I must have missed the original definition.

Resolution 4: Endorsement of the Washington Charter for Communicating Astronomy with the Public

I began to understand why, until now, no one had ever gone to the voting part of these meetings.

Resolution 5A: Definition of "planet"

Finally! I quickly read through the definition. Though confusingly worded and perhaps poorly thought through (not surprising, given that the final wording had probably been hammered out late in the night), the definition led to reasonable results. It even included a footnote that clearly stated, "The eight planets are: Mercury, Venus, Earth, Mars, Jupiter, Saturn, Uranus, and Neptune," and that Pluto and Xena, along with the asteroid Ceres, were to be called "dwarf planets," a term no one had ever heard before. The resolution was clear to point out that dwarf planets are not planets, which I found an odd use of the English language.

The first question from the press: "Dwarf planets are planets, right?"

No, I explained. The resolution was pretty clear. There are eight planets; dwarf planets, of which there might be hundreds, were clearly not planets.

But how could something be called a dwarf planet yet not be a planet? they wanted to know. A blue planet is a planet, right? A giant planet is still a planet. A dwarf tree is still a tree. How can a dwarf planet not be a planet?

Such is the beauty and frustration of definitions, I suppose. But I agreed that it seemed a poor choice, and an odd phrase to make up out of the blue. Something seemed suspicious. Still, the resolution was clear: There were only eight planets. If the astronomers voted yes on Resolution 5A, Pluto was clearly dead.

"But what about Resolution 5B?" someone asked.

I hadn't gotten around to reading that one yet. I turned to the screen.

Resolution 5B: Definition of Classical Planet

Huh? "Classical" planet? It was the Pluto escape clause! Resolution 5B simply changed the word *planets* in the previous resolution to *classical planets*. There would now be eight *classical* planets and four *dwarf* planets. With the quick addition of one word—*classical*—in front, dwarf and classical simply became different but equal subsets of the overall category of planets. Suddenly, dwarf planets *were* planets after all. The committee had indeed tried to sneak Pluto back in. The odd phrase *dwarf planet* had been invented in the previous resolution to allow the possibility that Pluto could rise from the underworld to live again.

Like the previous resolution, this definition was also muddled. Why "classical" planets? Shouldn't the phrase *classical planets* refer to those known in the classical world? In Greek and Roman times, there were seven planets: Mercury, Venus, Mars, Jupiter, and Saturn, and also the sun and the moon. Earth was not considered a planet, since it was the center of the universe. Uranus, discovered in 1781, and Neptune, discovered in 1846, were a few thousand years postclassical. Calling the largest eight planets "classical" made no sense at all.

I explained to the journalists in the room the now overly complex possibilities of the vote's outcomes and what the fate of Pluto would be in each case. Finally, the question came: "Do you think Pluto should be a planet?"

I sighed; it would have been thrilling to be considered the discoverer of a planet. "No," I responded. "Pluto should not be a planet. And neither should Xena. When Pluto was discovered in 1930, there was nothing else good to call it, but by now we know that it is one of many thousands of things in orbit out past Neptune. The vote today would rectify an understandable mistake made in 1930. Going from nine planets to eight planets would be scientific progress."

By 6:00 a.m. in California it was 3:00 p.m. in Prague, and the assembly was about to start. We were going to eavesdrop on the vote courtesy of a jumpy, low-resolution webcam broadcasting the event. I found the link for the webcast, clicked on it, and projected it onto the oversized screen behind me for everyone to see. It ended up filling an area of about one square foot. If you looked closely, you could see inch-high astronomers filing into the room.

Much of the next hour is a blur in my memory. After we watched an Austrian barbershop quartet there were nine hundred new members to be voted in and the first four resolutions to sit through. It would be a long morning. I muted the sound and opened the floor to questions, of which there were many. I can't remember a single one. On the tiny video, we could see people making speeches and raising yellow cards to vote on Barycentric Time. Someone finally brought me coffee. The first four resolutions quickly passed, with little discussion and without a single vote of no.

Finally the text for Resolution 5A appeared on the screen; we quickly unmuted the video stream and listened in. The once-secret committee, now beaten down by other astronomers, read and explained the resolution. The floor was opened for comment. One by one astronomers raised their hands and were passed a microphone. Here are some excerpts from the scientific debate of the learned astronomers:

Resolution 5A, Section 2 starts "a dwarf planet." Could you put dwarf planet in inverted commas, put quotation marks around dwarf planet? It is a definition. It should be in quotation marks.

The press assembled with me chuckled.

At the beginning of 5A we talk about planets and other bodies. That could be taken to include satellites. We didn't mean it to include satellites but it could be read to mean satellites.

The assembled press looked at me to see if this was significant. I shrugged my shoulders.

I suggest in part 3 of 5A where it says "all other objects" you insert "except satellites" and thank you to the people who suggested those points. I think they are a great improvement.

Chuckles all around.

The order in which we have the resolution printed is not the order in which some countries do business. In some countries you do the amendments first and then vote on the substantive resolution.

More laughs.

"Wait!" I said, quickly turning the volume down. "This comment is really important. This part is insidious. This is deliberate! 5B, which is an amendment to 5A, is voted on *after* 5A. 5A, which says Pluto is not a planet, will have general support, and then 5B will get snuck in to subvert the intentions of 5A. And no one seems to care."

But no one other than me seemed to grasp the enormity of the conspiracy at hand. Sure, perhaps I was a bit on the exhausted side at this point and inclined to believe that the secret committee had also conspired to assassinate Abraham Lincoln, Archduke Ferdinand, and Julius Caesar, but just because I was being paranoid didn't mean I was wrong.

I turned the volume back up, and we were back to punctuation:

The inverted commas look right when you see them, but you don't *speak* them. Could you not think of a new word which doesn't exist in the dictionary so that it doesn't have any baggage, and instead of calling it "dwarf planet," use some word, since it's an entirely new thing. . . . What you need is a new word rather than combination of old words; but a planet is a planet and so is a dwarf planet from a schoolmaster point of view.

I was feeling punchy and kept interjecting. "Yeah, he is right," I muttered. "'Dwarf planet' is a dumb phrase. For years we've called things like Pluto and Xena 'planetoids'—planetlike. That was a perfectly good word yesterday. But they're trying to be sneaky, they are. 'Dwarf planet' is dumb, but they need it so Pluto can become a planet with 5B."

The press at this point began to think that I was perhaps as crazy as all of the astronomers arguing over punctuation in Prague.

A question from the astronomical floor: "How does Charon fit?"

Right. At this minute there is confusion about Charon. If we pass 5A, Charon is not a planet. Right now I think there is confusion.

Someone else interjected: "It's a satellite! As long as it remains a satellite, it's out with this resolution."

Comment: "A point of clarification for me: Is a dwarf planet considered a planet?"

"That is Resolution 5B."

"In 5A a dwarf planet is not a planet?"

"Right."

In perhaps my favorite exchange of the very early morning, the question "Do I understand correctly that we are not anymore entitled to use the word 'planet' for planets around other stars?" elicited the response: "Are you referring to floaters, sir, or are you talking about extrasolar planets?"

Floaters? All I could think of were those little spots that you can sometimes see floating in your eye. I never heard the answer because I was at this point just shaking and shaking my head wondering how much longer this could possibly go on.

From a pedant: "Last Friday you mentioned we are not voting on the footnotes, but now you are referring to the footnotes. So are we voting on the footnotes or not?"

Response: "We were at one point trying to say that the footnotes are not part of the resolution. I think that position is not tenable; it is a stupid position. Therefore the footnotes are now part of the resolution."

Out of nowhere: "There is so much left in the resolution to common sense that I would propose to drop the entire resolution and leave Footnote One."

That was just about the best comment of the morning. The astronomer was right: The resolution that came up with a definition was so poorly written and vague that it would have been clearer to simply say what Footnote 1 said: The planets are Mercury, Venus, Earth, Mars, Jupiter, Saturn, Uranus, and Neptune. Everything else was just an attempt to explain why—and a poor attempt at that.

The commenting went on for another hour before, mercifully, someone called for a vote. Those in favor of Resolution 5A, which would create eight planets and an unspecified number of dwarf planets, were asked to hold a yellow voting card in the air. The room was filled with the color of the sun. There was no need to count. Resolution 5A passed with overwhelming support. Pluto was, correctly, no longer to be classified with the other eight planets. It was a moment that I never thought I would see in my lifetime.

The press in Pasadena were aghast and astounded and excited. They were ready to hit the "send" button to upload their stories.

"No no no, wait!" I told them. There was still Resolution 5B! This was where the conspiracy would happen! This was where the secret committee would subvert the will of the astronomical community! "Wait and watch!" I told them.

We watched. And then the most amazing thing happened. In the still-too-early fog of a not-enough-coffee morning in Pasadena, with the press watching astronomers half a world away, awaiting the secret sign to the pro-Pluto brotherhood to emerge to protect the god of the dead, I saw, instead, the moderator of the meeting stand up and make a few simple statements that put everything in precisely the right place. Where were the conspirators? Where were the daggers? Maybe I was in need of sleep.

Here is what she said:

> 5B involves inserting one word. Surely not a serious matter. However. For the benefit of non-astronomers present [but, really, isn't she doing this for the astronomers?], I want to do a bit of teaching, which demonstrates that resolutions are non-linear, and small changes have big effects. Excuse me while I dive under the table. [She pulls out a large beach ball, to represent planets, and a stuffed dog—Pluto!—to represent, well, Pluto.]
>
> At the moment, right now, having passed resolution 5A, we have planets, the eight that are named [*points to beach ball*], we have dwarf planets [*points to stuffed Pluto*],

and we have small astronomical bodies that are non-spherical. If we reject everything else this afternoon that is what will stand. If, however, we add the word "classical" to this group [beach ball], then we have adjective planets [beach ball], different adjective planets [stuffed dog], and it could be argued that what we are doing is creating an umbrella category called planets under which the classical planets and the dwarf planets fit. And if we do this then that [pulls out umbrella, puts beach ball and stuffed Pluto under it; audience erupts into applause] pertains.

"Who *is* that?" someone in the press asked me.

The speaker was Jocelyn Bell, who was widely considered to have deserved a Nobel Prize in 1974 for her discovery of pulsars. I didn't need to speak; I just smiled. No conspiracy was going to happen on her watch. Although I wasn't sure what the outcome would be, astronomers were going to decide based on knowing exactly what they were voting for.

Only two comments were allowed. The first, in favor of the pro-Pluto resolution, was from the member of the once-secret committee who had called and told me that the committee's original definition, now dead, had been assured of passing. Wearing a tie with planets on it, standing in front of the auditorium, he looked tense, angry, and maybe a little sad. He made his case:

Using the words "classical planets" is a compromise which allows more than one kind of planet in the universe. Yet advocates of the [eight-planet] model have refused this term. They will tell you why. Listen carefully. The word planet is being restricted to just one narrow point of view. Their restriction means that a dwarf planet is not a planet. It would be like saying a dwarf star is not a star. We can fix this. Will we have too many planets? Will we confuse the public? No. The distinction is as simple as an umbrella. Pluto is a planet, but it is in the dwarf planet category. So please pass 5B. The word planet must be shared.

I almost felt bad enough to want to give in. I didn't object to dwarf planets being considered planets, which was all he was talking about. But I did object to the other planets being termed "classical planets."

For the anti-Pluto side, a British astronomer stood up and spoke:

The key issue is the definition of the concept *planet*. This is a very important decision to be taken by the IAU; Resolution 5A is very close to the definition that was agreed by consensus at the meeting on Tuesday. There it was made clear that 3 distinct categories were being defined. Planet, dwarf planet, and small solar system body. The amendment [5B] proposes to insert the word "classical" in front of the word *planet*. It is inconsistent with the 1st paragraph of Resolution 5A. And it transforms 3 distinct categories into 2, planets and the rest, and that too has been made clear. In answer to the question, how many planets in the solar system? Resolution 5A gives the clear answer: 8. Resolution 5B implies at least 11 and soon several dozen. Both Pluto and Ceres become planets, and probably several main belt asteroids and several Kuiper belt objects as well. Resolution 5B not only removes a fundamental dynamical distinction for a planet, it is confusing and internally inconsistent. In my view it should be rejected.

Sadly, even though I was all in favor of rejecting the resolution, I found almost none of the arguments compelling. Who cared what the consensus was on Tuesday? The final vote was today! And really, if the concepts were significant, wasn't it more important to make sure to get them right than to worry about the precise wording of the resolution? Besides, would it matter if there were eleven or more planets? It wasn't the number that mattered, it was getting the concepts right. I realized that I wouldn't have minded if there were "major planets" and "dwarf planets" instead of "classical planets" and "dwarf planets." I guess my own version of pickiness was just as bad as anyone else's.

The vote was called. If the resolution passed, Pluto would be a planet again, and Xena would officially be part of the club. Chad, David, and I would be the only living discoverers of a planet in the world. At least for now. And still I didn't want it to happen.

"All in favor of the resolution?"

Astronomers in favor of 5B—in favor of repromoting Pluto—held up their yellow cards. There were many. The counting took a few minutes.

"Mister President, we report ninety-one votes in favor."

That didn't seem like enough, but I couldn't tell from the tiny webcast precisely how many astronomers were there in the auditorium.

"All opposed to the resolution?"

Astronomers opposed to 5B, who wanted to firmly cap the solar system at eight planets, held up their cards. A sea of yellow filled the auditorium, which immediately erupted in applause.

"I think, Mister President, a further count is not honestly needed."

"Then it's clear that Resolution 5B is not passed."

At that point it was final. And I said to the assembled press: "Pluto is dead."

The cameras whirred; correspondents talked into their microphones; on a screen on the other side of the room I could see myself on some local television station repeating, like an echo, "Pluto is dead."

Most of the remainder of the day was a blur of interviews, condolences, and congratulations. That afternoon I made my way to the studio of a radio station, where I was scheduled to be on a call-in program broadcast throughout Los Angeles. When I showed up at the studio, they told me that another astronomer would be calling in as a guest.

Great, I thought. Another guest would help me to stay focused and coherent.

When we went live on air, I suddenly realized that the astronomer was none other than the member of the once-secret planet-definition committee, live from Prague! It had been an even longer day for him than it had been for me.

He seemed tired, and he definitely didn't seem happy. He talked about how he thought the vote had done a disservice to astronomy. I said I thought astronomy had done a great service to the world.

He said that he was sad that no one would ever again be able to discover a new solar system planet under the current definition.

"You know," I said, over the radio to him half a world away, "when you tell me that no one will ever discover a planet again, I just take that as a challenge."

Over the course of the radio show, we both answered questions from callers. It was becoming clear that the idea that Pluto was no longer a planet was not going to be an easy sell.

Throughout the hour, the host collected suggestions for a new mnemonic for remembering the order of the planets. Some gave a slight modification of the previous standard—My Very Educated Mother Just Served Us Nine Pizzas—by turning "Nine Pizzas" into "Nachos" or into "Nothing," which was a bit funnier. But the best mnemonic, and the one that I still tell people to use to this day, sent in by an anonymous listener, sums up the feelings that would envelop much of the world over the next days, weeks, and months:

Mean Very Evil Men Just Shortened Up Nature.

Chapter Thirteen

DISCORD AND STRIFE

Keeping Pluto dead has taken a lot of work.

In the days, months, and years since the decision was made, I've been accosted on the street, cornered on airplanes, harangued by e-mail, with everyone wanting to know: Why did poor Pluto have to get the boot? What did Pluto ever do to *you*?

It is at these moments that I am most happy that astronomers ignored my initial advice to simply keep Pluto and add Xena and forget about a scientific definition. I am thrilled that astronomers instead chose to put a scientific foundation behind what most people think they mean when they say the word *planet*. They don't mean "everything the size of Pluto and larger," and they certainly don't mean "everything round." Instead, when people say "planet," they mean, I believe, "one of a small number of large important things in our solar system."

My job is just to explain the solar system as it actually is. People, I think, will then realize themselves that Pluto is not one of these large important things in our solar system.

Here is what I say to people:

Many astronomers, tired of the endless debates before and after the demotion of Pluto, will tell you that, in the end, none of this matters. Whether Pluto is a planet or not is simply a question of semantics. Definitions like this are unimportant, they will say. I, however, will tell you the opposite. The debate about whether or not Pluto is a planet is critical to our understanding of the solar system. It is not semantics. It is fundamental classification.

Classification is one of the first processes in understanding something scientifically. Whenever scientists are confronted with a new set of phenomena, they will inevitably, even subconsciously, begin to classify. As more and more things are discovered, the classifications will then be modified or revised or even discarded to better fit what is being observed and what they are trying to understand. Classification is the way that we take the infinite variability of the natural world and break it down into smaller chunks that we can ultimately understand.

So how should we classify the solar system? It's hard, because we are sitting in the middle of it and have known planets our whole lives. But let's try to do it from the perspective of someone who has never seen a planet before. Imagine that you are an alien who has lived your whole life on a spaceship traveling from a distant star to the sun. You don't know that planets exist. You don't even have a word for planet in your language. All you know is your spaceship and the stars you can see surrounding you. The sun—which originally looked like any other star—now gets brighter and brighter as your destination nears.

As you start to stare at and wonder about the sun, you suddenly notice that—wait!—the sun is not alone! You see that there is something tiny right next to it. You're excited beyond alien words. As your spaceship gets closer and you look even more carefully, you suddenly realize there are *two* tiny things next to the sun. No, three. No, four!

You have just found the things that we call Jupiter, Saturn, Uranus, and Neptune: the giant planets. From your perspective, still quite far from the solar system, they look tiny and so close to the sun as to be barely distinguishable. You don't have a word to describe them, so you make one up in your alien language: Itgsan.

You keep looking for a fifth Itgsan out beyond that fourth one you found, because it seems logical that there should be more, but even as your spaceship gets closer and closer to the system, you don't see anything out there. Trust me, I understand your disappointment.

Finally, as you get close and the four Itgsan get brighter and appear more distinguishable from the sun, you realize you were looking in the wrong place all along. There *are* other things next to the sun, but they are *inside* the first Itgsan, not outside. There are four of them, but they're much smaller than the first four things you found. So you come up with a new word. You call them Itrrarestles. You don't know it, but you've just found Mercury, Venus, Earth, and Mars.

For a very long time, as you keep getting closer, there is nothing new. Finally, when you're almost on top of the solar system, you realize that between the small Itrrarestles and the large Itgsan there is a band of millions and millions of tiny things going around the sun. And looking even more carefully, you see that outside the large Itgsan there is another band with even more. You call them something that I can't pronounce, but I call them the asteroid belt and the Kuiper belt.

Nowhere in that alien brain of yours would it be likely to occur to you to take one or two or even a few hundred of the things sitting in the Kuiper belt or in the asteroid belt and put them in the same category as the big things, the Itgsan and the Itrrarestles. Instead, you would quite rationally declare that the solar system was best classified by four major categories. And you would, I think, be correct.

The only thing wrong with our current classification of the solar system as a collection of eight planets and then a swarm of asteroids and a swarm of Kuiper belt objects is that it ignores the fundamental distinction between the terrestrial planets— Mercury, Venus, Earth, Mars—and the giant planets—Jupiter, Saturn, Uranus, Neptune. In the class on the formation of planetary systems that I teach at Caltech, I try to convince my students that, really, there are only *four* planets and that Mercury, Venus, Earth, and Mars shouldn't count. But even students who worry about their grades aren't willing to go that far. So, even though the aliens call them Itgsan and Itrrarestles, we'll lump them together and just call them all Tsapeln.

You can classify anything at all in many different possible ways. If you are studying birds, you might split them into land birds and seabirds; carnivorous birds and seedeaters; red birds, yellow birds, black birds, and brown birds. All of these distinctions can be important to you, depending on what it is you are studying about birds. If you are studying their mating habits, you might classify them in categories of monogamous and polygamous. If seasonal migration is your thing, you could classify them by those that stay put and those that fly south for the winter.

Things in the solar system can equally well be categorized in many different ways. Things with atmospheres. Things with moons. Things with life. Things with liquids. Things that are big. Things that are small. Things that are bright enough to see in the sky. Things that are so far away that only the biggest telescopes will ever see them. All of these are perfectly valid categories, and they might be of utmost importance to you if you study one specialized type of thing about the solar system. As with birds, your favorite solar system classification will depend on your interests.

Most people, though, don't have specialized interests in the solar system. The only classification scheme they will ever know is the word *planet*. They will know what a planet is and how many planets there are and what their names are. Their entire mental picture of what the solar system is, of how our local bit of the universe is put together, will be carried in the understanding of that simple word. The definition of the word *planet*, then, had better carry with it the most profound description of the solar system possible in a single word.

If you think of the solar system as a place consisting of eight planets—or, better, four terrestrial planets and four giant planets—and then a swarm of asteroids and a swarm of Kuiper belt objects, you have a profound description of the local universe around us. Understanding how such a solar system came to be is one of the major tasks of a wide range of modern astronomers. If, on the other hand, you think of the solar system as a place with large things that are round and smaller things that are not quite round, you have a relatively trivial description of the universe around us. There is nothing important to study here: We've known for hundreds of years that gravity pulls big things in space into the shape of a sphere.

. . .

Sometimes you don't even have to go through such extensive arguments. If you catch a person early enough, before the idea that Pluto deserves to be a planet has sunk in, you can teach things correctly from the start. Take Lilah, for example. Everywhere I went in the months following the IAU decision, people wanted to know if I thought Pluto had been treated fairly. Did I think Pluto was a planet? After a few weeks, I taught Lilah to answer for me.

"Lilah, is Pluto a planet?" I would ask, beginning our choreographed banter.

She would frown and shake her head.

"No no no no no no no."

As she got older the banter continued: "So what *is* Pluto, Lilah?"

"He's not a real dog. He's a dwarf dog."

My friends would laugh, and then invariably go out and buy Lilah Pluto toys. She has stuffed dogs, of course, but also a collection of nine-planet memorabilia. Early on she learned to figure out which one of the nine little circles on whatever picture she had was Pluto and then promptly declare, "Pluto is a dwarf dog." The continued laughs from that line were more reinforcement than I could possibly have given.

Another friend was worried how Lilah would react when she got older and discovered that I was a planet killer. "What will Lilah think," the friend said, "when she learns that Pluto is not a planet and that you are to blame?"

"I know what's going to happen," I replied. "In second grade or third grade, when she learns about planets she'll come home and say, 'Daddy, today we learned about the eight planets,' and I'll say, 'Lilah, did you know that when you were born we thought there were nine or even ten planets?' She'll look at me, shake her head, and say, 'You know, adults are *so stupid*.'"

. . .

Now that Xena, too, was officially called a dwarf planet, it finally got a real name. The possibilities were wide open, but Chad, David, and I had decided that because—at least in our minds—

Xena had been the tenth planet in good standing for an entire year, we wanted to give it a Greek or Roman name, like all of the other planets have. The problem was that there were very few left to go around. Back in the 1800s, when asteroids were first being discovered, they were, of course, called planets. And people wanted them to have Greek or Roman names, like the other planets. So they used up almost all of the major gods and goddesses and most of the minor ones, too. Every time we found a name we thought might be nice, we had to look it up in the databases of asteroid names to see if it had been used. Usually it had been. Finally, David wrote a quick computer program to correlate all asteroids with all names of Greek and Roman gods so we could see what—if anything—was left.

There wasn't much, and what there was was hardly recognizable. Obscure demigods of long-forgotten activities. Minor protectors of long-gone professions. But one name grabbed my attention. I remembered this name from my high school mythology readings, and I couldn't believe no one had used it before. Here was a major goddess with a fascinating backstory, overlooked in the solar system for two centuries. I quickly double-checked all of the asteroid databases. I double-checked that my mythological memory was correct. And then I sat down and wondered, for the first time since I had correctly predicted my sister's pregnancy, whether or not there was some sort of cosmic force governing the stars and planets and even the dwarf planets after all. Maybe there was some sort of fate that had kept this name free until now, the perfect time for it to be unveiled. Maybe there was no free will in any of this. That idea is, of course, crazy, but it's hard not to think crazy thoughts now and then.

I quickly e-mailed Chad and David, and we all agreed: the largest dwarf planet, temporarily nicknamed Xena, cause of the largest astronomical showdown in generations and the killer of Pluto, would henceforth be called Eris, after the Greek goddess of discord and strife.

I love the myth of Eris. As a perpetrator of discord and strife, she was not everyone's favorite goddess to have around, so when the human Peleus and the sea nymph Thetis decided to wed, they didn't invite her to the wedding. I understand their dilemma. Having gotten married myself, I know that there are always touchy issues involving the invite list. There are A lists and B lists and whole categories where you think, "Well, if I invite one person from this category, I should really invite *everyone* from this category," and then the bar tab gets out of control. If you find yourself having a wedding and are trying to decide whether or not to invite the goddess of discord and strife, my only recommendation to you is that if you decide *not* to invite her, make sure that she is not the *only* goddess who is not invited, which was the mistake Peleus and Thetis made.

The goddess of discord and strife doesn't take snubs lightly. She crashed the wedding anyway, and to cause, well, discord and strife, amid the guests she rolled in a golden apple on which she had inscribed *"Kallisti,"* meaning "to the fairest." As Eris had planned, all of the goddesses at the wedding got into a fight over who was the fairest and most deserving of the apple. They asked Zeus to decide. But Zeus, being no dummy, took the rather dim-witted mortal Paris, put him on the throne, and asked him to decide. The goddesses, being no dummies either, knew that they had best resort to bribery. Hera offered Paris domination over men. Athena offered Paris victory in battle. Aphrodite offered the love of the most beautiful woman in the world. Paris didn't have to think twice about that one and promptly handed Aphrodite the golden apple. Aphrodite then mentioned the fine print: The most beautiful woman in the world now did indeed love him, but she was married and living in Greece, and the Trojan Paris would have to go abduct her. He did, but the other Greeks didn't take it well. The decadelong Trojan War ensued.

I was sold, but I still had to name the moon of Eris. Gabrielle had been the obvious counterpart to Xena, but who went with Eris? I read through all of the literary mentions of Eris from the past. I pondered geographical considerations. I looked at family ties. I was in search of something very specific; I had a plan that I had told nobody. Again, fate intervened, and I found precisely what I was looking for. I sent the proposed name of the moon to the IAU, and I told no one.

At home that night, I told Diane all about Eris. She thought it was a fabulous name. "What about the moon?" she asked.

"It's a surprise," I said. "A surprise for you."

When the name Eris was announced in the press a few weeks later, many people who had been following closely got what they took to be the inside joke on the name of the moon. I had called the moon Dysnomia. Dysnomia was one of the children of Eris, and she was the daemon spirit of lawlessness. Xena on TV had been played by Lucy Lawless. People assumed that Dysnomia was a sly nod to that original nickname.

I was happy to take credit for the wordplay, but in reality it was an accident that I hadn't even noticed until someone pointed it out to me. I'll just chalk that up, once again, to cosmic fate.

On the day that the names were announced, I couldn't wait to get home to tell Diane.

"I named the moon for you," I told her.

"You named the moon Diane?" she asked.

I explained that since the name Diane had long ago been taken by an obscure asteroid, I had had to be subtle. When Jim Christy discovered Pluto's moon, he took the first syllable of

Charlene—his wife's name—and made a name out of it that's found in mythology: Charon. In searching for the perfect name for Eris's moon, I had looked for one that had the first syllable of Diane. Dysnomia is, admittedly, a bit clunkier than Charon, but there, in the first syllable, is my wife, Diane, whose family frequently calls her Di.

"Dysnomia is named for you," I said. "It's my present forever."

"Um, thanks, I think," said Diane.

After some contemplation, she added, "This doesn't excuse you from Christmas presents, you know."

A year earlier, when the existence of Xena was first announced, I had considered naming the potentially tenth planet after Lilah in some way. Diane had dissuaded me.

"What if we have a second child and you never find another planet?" she said.

That was a convincing argument.

I told her she should take the moon naming as a good sign: While it was possible that we might had a second child, there would be only one wife!

"Um, thanks, I think," said Diane, again.

. . .

Many people know about the Rose Parade, which winds through Pasadena every New Year's Day just as it did four days before the discovery of Eris in 2005. Less well known is the yearly alternative version of the Rose Parade called the Doo Dah Parade, which goes along some of the same main parade route as the Rose Parade. It attracts large crowds and features things such as marching toilets, a Doo Dah Queen (usually in drag), flying tortillas, and a Precision Grill team, cooking up barbecue along the way. In 2006, it also featured a New Orleans Jazz Funeral for

Pluto organized by some local astronomers with a sense of humor. The eight planets were each represented by a costumed astronomer with a large cardboard name tag hanging around his or her neck. They carried Pluto in a casket to sounds of New Orleans jazz. The astronomers invited me to participate and gave me a cardboard name tag that read: "Mike Brown: Pluto Killer." I had agreed to march in the parade on one condition: that Eris also be invited. Eris, played by Lilah, was pushed in a stroller down the parade route by her father.

Like most marchers in the Doo Dah Parade, we got quizzical looks, a few claps, a smattering of boos, and a lot of tortillas thrown at us. I spent most of my time trying to make sure Lilah didn't pick them up and eat them. But still, Pluto was dead, and it was good to participate in its burial.

. . .

But not everyone was ready to bury Pluto just yet.

On the very day of the vote to demote Pluto and Eris, a few astronomers began collecting signatures protesting the details of the IAU decision. They issued a simple statement:

> We, as planetary scientists and astronomers, do not agree with the IAU's definition of a planet, nor will we use it. A better definition is needed.

It's hard to argue with that statement. As much as I am proud of the astronomers who had the guts to go against emotional sentiment and to remake the solar system correctly, the actual definition by which they did it is pretty clunky. In fact, I won't use it, either.

In the question-and-answer session of a recent talk I gave at Sarah Lawrence College, a very agitated young woman raised her hand and began to read from notes: "In the IAU definition of the word 'planet' it says you have to be three things to be a planet . . ."

"Wait, wait, wait," I said. "Before you even start, let me tell you why you should *never* think about the IAU definition of the word 'planet.'"

In the entire field of astronomy, there is no word other than *planet* that has a precise, lawyerly definition, in which certain criteria are specifically enumerated. Why does *planet* have such a definition but *star, galaxy,* and *giant molecular cloud* do not? Because in astronomy, as in most sciences, scientists work by concepts rather than by definitions. The concept of a star is clear; a star is a collection of gas with fusion reactions in the interior giving off energy. A galaxy is a large, bound collection of stars. A giant molecular cloud is a giant cloud of molecules. The concept of a planet—in the eight-planet solar system—is equally simple to state. A planet is one of a small number of bodies that dominates a planetary system. That is a concept, not a definition. How would you write that down in a precise definition?

I wouldn't. Once you write down a definition with lawyerly precision, you get the lawyers involved in deciding whether or not your objects are planets. Astronomers work in concepts. We rarely call in the attorneys for adjudication.

The young woman in the audience was not satisfied.

"You can't just dismiss the definition. The definition is the reason that Pluto is no longer a planet!"

I tried to explain to her that the *concept,* not the definition, is the reason that Pluto is not a planet. The definition was simply a poor attempt at codifying the concept.

She went on: "But by part three of the definition even Jupiter is not a planet!"

The young woman could probably make a reasonable case in

court for her strict reading of the definition. But when the case was appealed to the Supreme Court—and it certainly would be—some justices might try to discern the original intent of the definers. I am certain that it was not anyone's intent to exclude Jupiter from being a planet. The original intent was simply an attempt to describe the eight-planet solar system. The case for a strict reading of the definition would ultimately be tossed out. And then, if the justices were wise, they would also toss out the definition altogether. We're better off without one. Pluto is not a planet not because it fails to meet the three-headed criteria laid out by the IAU. Pluto is not a planet because the criteria were written to try to explain the concept that Pluto is not a planet.

. . .

But the astronomers who organized the petition saying that they would never use the IAU definition were not quibbling over the logic of having a definition in the first place. They wanted the eight-planet solar system overturned. They wanted Pluto resurrected. While most of the rest of the astronomical world has acknowledged the reasonableness of the decision and moved on, a small group is continuing to try to have Pluto make a comeback.

Over the months and years, their arguments have changed, in the attempt to get some traction. At first, they took a line straight from the people trying to get creationism taught alongside evolution in schools: "Teach the controversy!" they said. Then they argued that the IAU decision was undemocratic because many of the members of the IAU had not been there that day to vote. The complaint is true, but the implication that the outcome would have been different is quite a stretch. Sometimes the argument is that only *planetary* astronomers are qualified to make the decision—again, as if that would make a difference. In my unscientific poll of seven professors of planetary science who happen to work on the same floor as I do, all seven thought that eight planets make the most sense.

Particularly amusing to me was the complaint about the phrase *dwarf planet*. By the simple rules of grammar, a dwarf planet is a planet, they would say. The fact that the IAU would say that a dwarf planet is not a planet demonstrates that the entire decision must be wrong. What no one making these arguments remembers—or admits to remembering—is that the only people who liked the phrase *dwarf planet* at first were the ones who hoped that it would save Pluto when the other planets were renamed "classical planets." Yet Resolution 5B was a specific vote on this issue, and it clearly stated that dwarf planets are not planets—just as Matchbox cars are not cars, stuffed animals are not animals, and chocolate bunnies are not bunnies. I don't particularly like the phrase *dwarf planet,* either, but it is serviceable.

I've heard the argument that the definition is unworkable because it is inconsistent with the rest of astronomy. Nowhere else in astronomy, some say, do you classify an object by its relationship to its neighbors instead of by its own individual properties. Therefore the only definition that makes sense is that all round things are planets, regardless of where they find themselves. Well, not all round things are planets, just round things that orbit stars. And what if the round thing orbits another round thing? Well, then it is a moon, of course. But but but, doesn't that violate the rule that things aren't defined in relationship to other things? Well, yes, but that's just common sense, they would say. Okay. Got it.

The biggest boost to the attempted resurrection of Pluto came from Eris itself. When we first measured the size of Eris we found that it was 1,490 miles across, or about 5 percent larger than Pluto, but the uncertainty in that measurement was as much as 4 percent, meaning that, according to our analysis, Eris could be anywhere between 1,430 and 1,550 miles across (while Pluto was thought to be a piddling 1,400 miles across). That 120 miles of uncertainty is relatively small, but since we are talking about the size of the largest thing found in the solar system in 150 years, it would be nice to have a more accurate answer. In order to pin down a more precise size, astronomers in France spent years tracing the path of Eris through the sky and predicting its future position. Finally, they determined that on one night in late November 2010, Eris would pass directly in front of one relatively bright star. When that night arrived, astronomers around the entire dark half of the globe lay in wait, staring at the star, watching to see if anything would happen. I was watching from a little telescope at Palomar Observatory and saw nothing but the clouds that had been around all November long. But astronomers in Chile had better luck: at precisely the predicted moment, the star abruptly disappeared. About 76 seconds later, it reappeared, just as abruptly. The amount of time that the star had not been visible told them precisely how large Eris really was.

I bit off all of my fingernails waiting to hear the results. It wasn't that I was nervous for the sake of Eris—Eris would be fine no matter what the result—I was just anxious for myself. I had measured the size of Eris and had declared it to be between 1,430 and 1,550 miles in diameter. I didn't *think* that I was wrong, but it was an exceedingly difficult measurement, just at the edge of what the Hubble Space Telescope was capable of. If I had made any mistakes or errors or miscalculations everyone was soon going to know. It's one of the joys of science: If you do something wrong, you will eventually be found out. It's quite a motivator to not make too many mistakes.

I needn't have massacred my fingernails, though. When I

finally got the word from Chile and from France, Eris scraped in on the small side of my predicted range. They found it to be 1,445 miles across: three percent larger than Pluto, just as we had calculated.

Scientifically, the close match in size between Eris and Pluto is astonishing. For comparison, the diameters of the two largest asteroids, Ceres and Vesta, are nearly a factor of two different from each other, equal to the difference between driving from Los Angeles to San Francisco and going from Los Angeles all the way to the Oregon border. Eris and Pluto, on the other hand, are much bigger—the driving distance between the Washington, D.C., area and Denver—but the size difference between the two is the difference between starting your trip in Washington, D.C., itself or in Baltimore instead. Even more astonishing to me, though, is that even though Eris and Pluto appear nearly identical on the outside, on the inside they are shockingly different. Eris is still 27 percent more massive than Pluto—that measurement didn't change. If Eris and Pluto were made of the same stuff on the inside, Eris should have a volume 27 percent larger than that of Pluto, which would mean that its diameter would have to be 9 percent larger than Pluto's. When we first calculated that Eris was 5 percent larger than Pluto with an uncertainty of 4 percent, we were fairly convinced that we knew the real answer: It made sense that Eris must actually be 9 percent larger, and that Pluto and Eris were made out of the same materials. We were wrong. Since Pluto and Eris are nearly identical in size, the interior of Eris must be made out of heavier material. In this part of the solar system, the insides of bodies are almost entirely made up of rock and ice. Eris weighs so much that it must be almost entirely rock, while Pluto has significant quantities of frozen water hidden inside it.

Differences like these might seem small, but they contain

enormous clues to how the entire solar system formed, billions of years ago. Did large bodies like Pluto and Eris form in different places? Have they had dramatically different histories? Answer these questions and you are one step closer to unraveling the mysteries of the solar system. And then please tell me, because I still haven't figured it out.

I knew that the press wouldn't be all that interested in the details of rock and ice, but I was certain they would be eager to learn the true size of Eris. I couldn't wait to read the headlines. I assumed they would say something like: "New measurements confirm that Mike Brown was precisely correct!" Or, perhaps more realistically: "Eris and Pluto are the same size and therefore very different!" But I was wrong. On January 10, 2010, the headline in *The New York Times* read "The War of the Worlds, Round 2," and the story suggested that in the battle of dominance between Pluto and Eris, Pluto was now winning. Other headlines that began rolling in all contained variants of the same theme: "Astronomers erred; Eris is now known to be smaller than Pluto!" and, most strangely, "Pluto should be a planet again!"

Wait, what?

Why would anyone say that Eris is smaller than Pluto when it is three percent bigger? Reading carefully, I realized that something very strange was happening. In the six years since the discovery of Eris, the generally accepted size of Pluto had crept upward while almost no one was paying attention. Pluto had secretly been expanding! It's not as if there were new measurements that showed that Pluto was larger, it's just that when you asked people who studied Pluto how big it really was, the answer that they gave tended to be a little higher than before. At the time of the discovery of Eris, that generally accepted answer was around 1,400 miles. Now people were putting Pluto's size at as large as 1,450 or even 1,470 miles. If Pluto were this size, it would be larger than Eris after all.

So how large is Pluto? The honest but embarrassing answer is that we don't actually know as precisely as we often pretend. The size of Pluto has been measured by different people using different techniques and the answer has ranged from as small as 1,400 to as large as 1,470 miles. The shift in the accepted size of Pluto over the past few years has mostly been a shift in whose measurements people believe. Of course, the scientifically precise way to answer the question is not simply to see whom people believe most on any one day, but to carefully examine all the different measurements that have been done and try to understand what they are really telling us. What they tell me is that something fishy is happening on Pluto, and the most likely culprit for this fishiness is an atmosphere on Pluto that is obscuring the surface and confusing the answer. In light of this confusion, I have to admit that I don't know which measurement is most accurate.

Getting the size of Pluto right is critical, though, because, as the headlines implied, if Pluto is bigger than Eris then Pluto should be a planet again, right? Of course not. Pluto was not demoted simply because it was no longer the largest known object beyond Neptune but because it was one of many, many such small objects beyond Neptune. The fact that it might still be the largest gives it some bragging rights at the next dwarf planet convention, but being the largest known thing beyond Neptune doesn't get you an invitation to the planet ball.

You could certainly argue that if nothing larger than Pluto had ever been discovered Pluto might have kept its status forever, and you might even be right. Eris gave astronomers the final impetus to straighten out the solar system and correct a mistake from 1930, and without its prompting they might well have just left it alone. But even if Eris had never been discovered, demoting Pluto would still have been the right thing to do.

So which one really is bigger? My gut feeling is that the smaller end of the Pluto numbers is correct, but only because I think the techniques used for those measurements are less susceptible to fishy atmospheric effects. But it is likely that we won't know the real size of Pluto until the *New Horizons* spacecraft flies by in 2015 and gets a much closer look. It is a bit amusing to contemplate that one of the most memorable discoveries of the mission to Pluto might be that Pluto is, after all, smaller than Eris.

. . .

A few weeks after Xena became Eris, I received a note from a friend:

The Spanish are trying to steal Santa again.

The Spanish? I hadn't thought much about them in the preceeding eighteen months and certainly hadn't heard anything from them. By this time, I was almost able to laugh about the whole incident.

But they were really back.

After the IAU decided to call round things dwarf planets, Easterbunny and Santa were eligible for real names, too, and the Spanish astronomers quickly submitted a name for Santa—because, of course, the discoverers get to name their discovery.

Rumor had it that the IAU was going to act swiftly, so Chad, David, and I quickly consulted and came up with our own name: Haumea, after the Hawaiian goddess of childbirth. Like the name Eris, the name Haumea seemed almost custom made for this object. The goddess Haumea gave birth to her many children by breaking them off from parts of her body. Santa the dwarf planet also had many children throughout the solar system that had broken off from its body. It seemed a perfect fit. And whatever the name was, it *definitely* should not be whatever the Spanish astronomers were submitting!

I wrote an impassioned letter to the various committees of the IAU proposing the name Haumea and also proposing names for two moons: Hi'iaka, the patron goddess of the Big Island, and Namaka, a water spirit, both daughters of Haumea. In the letter, I laid out—once again—all that had transpired. And then I explained why it was important that the IAU choose wisely which name to use. There was no question that eighteen months earlier someone had done something unseemly. If the Spanish astronomers had fraudulently claimed discovery of something that they had never actually discovered, it would be appropriate for the IAU to condemn such a thing. If, on the other hand, their discovery was legitimate, they should be exonerated and I should be censured for making a spectacularly damaging wrongful accusation. Through choosing a name, the IAU would be officially choosing a side. I thought that the members of the IAU would not want to take sides and would instead pick a name themselves. I urged them not to cop out. No one else had the authority to render any type of meaningful verdict.

I sent in my letter and waited to see what would happen.

And I waited.

Lilah's second birthday came around.

And I waited.

Lilah's third birthday came around.

I was hoping that the IAU was taking its job quite seriously and had launched a multiyear investigation into what had happened. Apparently not. My inside sources tell me that nothing happened that entire time. Finally I got a tip that making a decision about Santa was just too hard and that perhaps we should try giving a name to Easterbunny as a way of getting things restarted.

Ah, Easterbunny. I had been thinking about it now for years. The names Sedna and Orcus (another large Kuiper belt object that we had turned up) had fit the characteristics of the objects' orbits, and the names Eris and Haumea had practically fallen out of the sky at us. Even Quaoar, we felt, was a nice tribute to local mythology.

But what about Easterbunny? Unlike Santa, which has so many interesting characteristics that there were many possible names, Easterbunny has no obvious hook. Its surface is covered with large amounts of almost pure methane ice, a consequence of the fact that it is just a little smaller than Pluto and lacks enough gravity to hold a substantial nitrogen atmosphere, which is scientifically fascinating and all (really, it is) but not easily relatable to terrestrial mythology. For a while I was working on coming up with a name related to the oracles at Delphi: Some people interpret the reported trancelike state of the oracles to be related to natural gas (methane) seeping out of the earth there. After some thought, I decided this theme was just dumb. Strike one.

I spent some time considering Easter- and equinox-related myths, as a tribute to the time of discovery. I was quite excited to learn about the pagan Eostre (or Oestre, or Oster, or many other spellings) after whom Easter is named, until I later realized that this mythology is perhaps itself mythological and, more important, that an asteroid had already been named after this goddess hundreds of years ago. Strike two.

Finally, in mythological desperation, I considered rabbit gods, of which there are many. Native American lore is full of hares, but they usually have names such as Hare or, better, Big Rabbit. I considered Manabozho, an Algonquin rabbit trickster god, but I must admit, perhaps superficially, that the "bozo" part at the end was a turnoff. There are many other names of rabbit gods, but the names just didn't speak to me. Strike three.

These initial attempts had all happened long ago, and I had given up, figuring I would wait for the IAU decision on Santa. But now, with some prodding, I got back to work.

Suddenly, it dawned on me: There was a potentially interesting small island in the South Pacific that I hadn't looked into before. I wasn't familiar with the mythology of the island, so I had to look it up, and I found Makemake (pronounced Hawaiian style as "mah-kay mah-kay"), the chief god, the creator of humanity, and the god of fertility. I had discovered Easterbunny during the time that Diane was pregnant with Lilah. Easterbunny was the last of these discoveries. I have the distinct memory of feeling a fertile abundance pouring out of the entire universe during that time. Easterbunny was part of that. Easterbunny would be Makemake, the fertility god of the island of Rapa Nui.

Rapa Nui was first visited by Europeans on Easter Sunday, 1722, precisely 283 years before the discovery of the Kuiper belt object now known as Makemake. Because of this first visit, the island is known in Spanish (it is a territory of Chile) as Isla de Pascua, but around here, it is far better known by its English name of Easter Island.

. . .

The name Makemake was accepted quite quickly and with a modest fanfare by the IAU; as predicted, a decision on Santa was soon rendered, only two years after the initial proposals had been submitted. This time there was no fanfare, no press release, no official pronouncements. The name just appeared on the official IAU list of names one day as Haumea. Three years after the

Spanish astronomers either did or did not fraudulently steal our discovery, we were officially vindicated by the IAU, which accepted our name, signaling that we appropriately deserved the credit.

Sort of.

On the IAU's list, next to the newly added name Haumea, in the space reserved for the name of the discoverers, is a big blank spot. Haumea, unique among all objects in the outer solar system, has no discoverer. It simply exists.

Oddly, though, for an object that no one discovered, it does have *place* of discovery listed. While the name of the object is Hawaiian, based on a proposal by astronomers from California, Haumea was officially discovered at a small telescope in Spain. By nobody.

What does any of this mean, officially? Mostly, I think, that the IAU didn't try too hard to figure anything out. Probably the majority of whatever committee was voting thought my version of the story was the most plausible, but there were enough dissenters that a decision was made to soften the pronouncement by listing no discoverer and by backhandedly acknowledging the Spanish claim.

I am disappointed that they made no real effort to figure out what happened, at least as far as I can tell. No one ever asked me anything or requested extra information from me. I suspect the same is true of the Spanish side. In the end, this is as good as it will get. I will never know for sure what actually took place in those two days before the Spanish astronomers announced their discovery.

· · ·

I still haven't drunk the celebratory champagne. The friend with whom I made the five-year bet on the foggy night at Palomar

Observatory had generously given me a five-day extension, and Eris fit all of the characteristics that she and I had decided a planet must meet. She happily delivered the champagne the next time she was in town. In the end, though, Eris was not the tenth planet; it was instead the killer of the ninth. Champagne doesn't make a good funeral drink.

Those five champagne bottles sit on my shelf still. I look at them every once in a while and wonder if the time will ever come to pop the corks. I'm still looking for planets, but the bar is now much higher. Anything new that wants to be called a planet needs to be a significant presence in our solar system, and I am not certain that there are any more hiding in the sky. But I keep going. Someday, I hope, I'll be sitting in my office looking at pictures of the sky from the night before, and there on the screen will be something farther away than I've ever seen before, something big, maybe the size of Mars, maybe the size of the earth—something significant. And I'll know. And, as I did years earlier, I'll immediately pick up the phone and call Diane. "Guess what?" I'll say. "I just found the ninth planet." And—once again—the solar system will never be the same.

JUPITER MOVES

It takes some time for a kid to figure out that her parents have a separate existence that takes place when she's not around. By the time she was about three, whenever I was gone for a few days Lilah began getting immensely curious about where I was. That place would become a fabled land that she invoked when playing with stuffed animals or making up stories. Taiwan, to which I went one week during her third year and which she can now pick out on any globe, remains perhaps her favorite spot in the world. At one point during her third summer, she had named all of the corners of the swimming pool after different places, and she would cling to my back and direct me where to go next.

"Daddy, I want to go to Chicago."

Swim, swim, swim.

"Daddy, Daddy, Berlin!"

Stroke, stroke, stroke.

"Boston."

Glide, glide, glide.

"Daddy, Daddy, I want to go all the way to Taiwan!"

Taiwan, which she knew to be an island, required momentarily going underwater before emerging on the other side of the Pacific Ocean.

"Now back to Pasadena, California!" which was code for "Let's get out and see if Mommy will bring out some snacks."

Eventually she started figuring out why I would periodically disappear.

"Are you going to go talk about planets?"

And the answer, invariably, was yes.

Lilah loves planets. Other than the occasional dwarf-dog joke, I have never particularly pushed planets on her, at least not any harder than I push them on everyone else. Yes, I point out planets in the sky to her every time we go outside at night, but I do that to everyone. Beginning in that summer of her third birthday, Lilah had been particularly mesmerized by Jupiter. Every night for a few months, it was high in the evening sky— one of the first things to pop out of the murky twilight and reveal itself night after night. Back in the summer, she made sure we went outside right at her bedtime, when it was just barely dark enough to make out Jupiter, so she could say good night to it. As the summer changed to fall and then winter, it would already be dark as we were driving home, and for her, the highlight of the drive was always the moment after we'd climbed the little hill to our neighborhood and we had taken the final left-hand turn to point west; Jupiter suddenly would appear in her window, high enough in the sky to be seen even from the depths of her child car seat.

By late fall, though, Jupiter was no longer the king of the evening skies. Venus had crept up into the twilight to start to steal the show from Jupiter. Or at least, in Lilah's view, to share the show. She went from having only one planet to now having two planets to say good night to every night.

Lilah sees planets everywhere. You never quite realize—until you have an obsessed three-year-old—how prevalent images of planets are in everyday life. She's got them on her lunch box (a gift from friends of mine, of course); she sees pictures in magazines and catalogs; she sees mobiles and puzzles at stores. I would tend to just walk by them without noticing, but she always runs up—"Daddy, Daddy, *look*!" She always quickly picks out Jupiter (the big one) and, of course, Saturn, with the rings. She recognizes the blue-and-green look of Earth. And she gets Venus right more often than I think any three-year-old should.

One night, after a long cloudy spell when we couldn't see the planets at night, Lilah looked up at the sky, startled. "Daddy, Daddy, *look*! Jupiter *moved*!" And she was right. Venus and Jupiter had been slowly edging closer to each other over the past few weeks, but you wouldn't have noticed it unless you were watching closely. Now they were suddenly so close that even a three-year-old could see that something had changed.

Lilah's pointing out to me that Jupiter moved was—for me—the pinnacle of planetary charm. While most kids and adults can name the planets and point out pictures, almost nobody notices the real things, even when they are blazing in the evening sky. Planets are not just things that spacecraft visit and beam back pictures from. They're not just abstractions to put on lunch boxes. They are really there, night after night after night, doing what planets do: moving; wandering.

A few nights later, the show got even better. A tiny sliver of a moon appeared low in the early-evening sky and began working its way toward Jupiter and Venus. Lilah and I are moon watchers. And we both know that after appearing as a tiny sliver at sunset, the moon gets bigger and moves east night after night in the evening sky. Based on how far the moon was from Venus and Jupiter, it was clear that in just two nights the moon would be packed tight right next to Jupiter and Venus. It would be a spectacular sight, with the three brightest objects ever visible in the night sky in an unmistakable grouping in the southwest just after sunset.

The night of the triple conjunction, I was on a long flight across the country. As I was packing my bags that morning, Lilah had sadly asked, "Daddy, are you going away to go talk about planets?"

I was. But I didn't want talking about planets to make me miss seeing planets. I knew I was touching down at night in Florida long after Jupiter and Venus and the moon would have set, but I was careful to pick a window seat on the south side of the airplane so I could watch the show from the air. And the sight of the moon and Jupiter and Venus shining in a tight triangle over and behind the wing was as spectacular as Lilah and I knew that it would be. Though it was night in Florida, it would still be a beautiful late twilight in California. I quickly called home and told Lilah all about the view from 30,000 feet and told her to go outside right now and—*look!* She would see all of our favorite planets.

The tight-packed group of lights low in the early evening sky was the sort of sight that makes even non–night sky watchers suddenly look up and wonder. A few people would even think to look the next night, I suspected, to see if the sight was still there. They would notice that the moon had already moved farther east and gotten a little bigger, and they would see that the two other bright lights—Jupiter and Venus—were in slightly different spots than just one night earlier. Maybe then a person or two would be hooked. Maybe they would follow the moon's movement for the next week as it grew to full, watching as Jupiter appeared lower night after night, eventually leaving Venus alone in the sky. It would be a show worth following. I knew Lilah and I would watch it. Even when we were continents apart, we'd always be looking for the things that moved in the sky.

ACKNOWLEDGMENTS

This book would never have been possible without the contributions of many people involved in the research and events described here. I would like to especially thank Jean Mueller and Kevin Rykoski for their early encouragement of and help with the search for large objects in the outer solar system, and Chad Trujillo and David Rabinowitz for many years of hard work and foresight into what might be out there and how to find it. Brian Marsden was always a voice of wisdom and kindness in the otherwise arcane world of solar system politics. My students throughout this period, Antonin Bouchez, Adam Burgasser, Lindsey Malcolm, Kris Barkume, Emily Schaller, Darin Ragozzine, and Meg Schwamb—now Drs. Bouchez, Burgasser, Malcolm, Barkume, Schaller, Ragozzine, and Schwamb—all provided fresh eyes and minds that aided many of the scientific insights described here.

While the research and discoveries were key, the book itself might not have ever been begun without encouragement from Heather Schroder on an early abortive version, and then a jump start from my agents, Caroline Greeven and Marc Gerald, who finally set me to work. Cindy Spiegel took the initial manuscript and found a way to make small changes with big impacts and graciously laughed at me when I told her I was nervous to meet *real* writers. Brad Abernethy provided wonderful editorial advice and encouragement on an early draft, and explained to me that words mean what we think they mean when you say them. Emily Schaller, though mentioned above in her doctoral capacity, also deserves my deepest gratitude for reading every version of every chapter and always providing exactly the right combination of advice, criticism, and encouragement.

I regret that my father, Tom Brown, didn't live to see most of the time period written about here, but he was nonetheless instrumental in instilling in me my love for space, science, and living on boats. My mother, Barbara Staggs, has always been my biggest fan, no matter what the arena, and my stepfather, Willie Staggs, my brother, Andy Brown, and my sister, Cammy Thornton, have always kindly tolerated this fact and provided balance, for which I am grateful.

Finally, I have to thank Diane and Lilah, who are the reason for the book and also the ones who allowed it to happen, by letting me mentally slip away on nights and on weekends to write the stories of us, and who continuously allow those stories to go on.

INDEX

Entries in *italics* refer to illustrations.